T0178341

Understanding NEC3: Engineering and Construction Short Contract

As usage of the NEC (formerly the New Engineering Contract) family of contracts continues to grow worldwide, so does the importance of understanding its clauses and nuances to everyone working in the built environment. This set of contracts, currently in its third edition, is different from others in concept as well as format, so users may well find themselves needing a helping hand along the way.

Understanding NEC3: Engineering and Construction Short Contract uses plain English to lead the reader through the contract's key features, including:

- the use of early warnings
- programme provisions
- payment
- compensation events
- preparing and assessing tenders.

Common problems are signalled to the reader throughout, and the correct way of reading each clause explained. In addition, the things to consider when deciding between the Engineering and Construction Short Contract (ECSC) and the longer Engineering and Construction Contract are discussed in detail.

Written for professionals without a legal background, by a practising construction contract consultant, this handbook is the most straightforward, balanced and practical guide to the NEC3 ECSC available – an ideal companion for Employers, Contractors, Project Managers, Supervisors, Engineers, Architects, Quantity Surveyors, Subcontractors and anyone else interested in working successfully with the NEC3 ECSC.

Kelvin Hughes spent 18 years in commercial management with major contractors, then the past 21 years as a consultant, including a four-year senior lectureship at the University of Glamorgan (now the University of South Wales), UK. He has been a leading authority on the NEC since 1996, was Secretary of the NEC Users' Group for ten years and has run over 1,400 NEC-based training courses.

Understanding NEC3: Engineering and Construction Short Contract

A practical handbook

Kelvin Hughes

Routledge
Taylor & Francis Group

LONDON AND NEW YORK

First published 2014
by Routledge
2 Park Square, Milton Park, Abingdon, Oxon, OX14 4RN

and by Routledge
711 Third Avenue, New York, NY 10017

Routledge is an imprint of the Taylor & Francis Group, an informa business

© 2014 Kelvin Hughes

British Library Cataloguing in Publication Data
A catalogue record for this book is available from the British Library

Library of Congress Cataloging-in-Publication Data
    Hughes, Kelvin (Engineering consultant)
    Understanding NEC3, engineering and construction short contract :
    a practical handbook / Kelvin Hughes.
    pages cm
    Includes bibliographical references and index.
    1. Civil engineering contracts—Great Britain.
    2. Construction contracts—Great Britain. 3. NEC Contracts. I. Title.
    KD1641.H832 2014
    343.4107'869—dc23
    2013043318

ISBN: 978–0–415–65711–2 (pbk)
ISBN: 978–1–315–78031–3 (ebk)

Typeset in Goudy
by Swales & Willis Ltd, Exeter, Devon

Contents

Figures

Preface

I have been involved with the NEC family of contracts since 1995 and in that time have advised on numerous projects using the contracts, and carried out over 1,400 NEC-based training courses. I was also Secretary of the NEC Users' Group from 1996 to 2006, providing support to users of the contracts, including seminars and workshops, during which time I also ran the Users' Group Helpline, answering queries on the NEC contract from members of the Group.

During this involvement with the NEC3 contracts, particularly with the Engineering and Construction Contract (ECC), I have always felt there was a need for practical manuals on the contracts and their use, including worked examples and illustrations; for example, what does an activity schedule look like? How do time risk allowances and float work? How do you assess compensation events for omissions of work? These are not covered either by the Guidance Notes or by other textbooks.

The Guidance Notes that accompany the contract were written by the contract drafters and, whilst they are comprehensive and well written, I have always felt that they do not adequately act as a practical handbook. Various books that are available have been written by 'NEC observers', with those being published by Thomas Telford often tending to promote the contract, rather than helping people to use it.

The Thomas Telford publications also tend to be passed in front of the NEC Panel, partially of course so that they are technically correct, but also, I feel, to ensure that nothing is said that could damage the reputation of NEC.

It is certainly not my intention to damage the reputation of NEC3, which I believe is an excellent family of contracts, well written by an extremely knowledgeable and professional Panel, many of whom I know personally and for whom I hold a great admiration.

This book is intended to fill the gap where the Guidance Notes leave off, and cover issues that may not have been considered or covered within the official publications.

It is intended to be of benefit to professionals who are actually using the contract, but also to students who need some awareness of the contract as part of their studies.

The Engineering and Construction Short Contract was formally launched at the Institution of Civil Engineers in London on 30 July 1999, having previously been trialled in its prepublication 'pilot' form by five major corporate users.

This book picks up from the second edition, launched in June 2005 with the NEC3 family of contracts, and incorporates the following reprints and amendments to the contract to bring it up to the present day:

- reprinted 2007, 2008, 2009, 2010, 2011
- reprinted with amendments 2013.

References to clause numbers in the book relate to the April 2013 reprint, though the clause numbering is not dissimilar to the original July 1999 and subsequent NEC3 June 2005 versions.

Whilst a number of practitioners still use the NEC 1st Edition Engineering and Construction Short Contract, the book is primarily aimed at giving guidance to NEC3 users, though the structure and contents of the contracts are very similar and to that end much of the advice given in this book is of use to all users.

Note that throughout the book I have made reference to the Engineering and Construction Short Contract using the initials 'ECSC', the Engineering and Construction Short Subcontract using the initials 'ECSS' and to the Engineering and Construction Contract using the initials 'ECC'.

As I have always viewed the NEC3 contracts as a manual of good practice as well as a contract, particularly in terms of the use of its main and secondary options, risk management through the early-warning system, clear requirement for a programme and a disciplined procedure for change management, I have deliberately included within each chapter an overview of the subject area and how contracts deal with the issues, as well as detailing the NEC3 provisions.

Readers of my two previous books, *Understanding the NEC3 ECC Contract: A Practical Handbook* and *Understanding NEC3: Professional Services Contract: A Practical Handbook* will see some duplication from those books when reading this one, particularly when giving general advice, and this was deliberate, my intention being that each book should stand on its own, rather than having to read the first book to understand the second and now the third book properly.

Readers may note the absence of case law within the text of the book. This is a deliberate policy on my part, for three reasons.

First, I am not a lawyer, my background being in senior commercial positions within major building contractors, so I felt, and readers may concur, that I was unqualified to quote and to attempt a detailed commentary on any case law.

Second, there has actually been very little case law on the NEC contracts since they were first launched.

Third, and probably most importantly, as a contracts consultant with significant overseas experience of all contracts, including the NEC, it was always my intention that the book should attract an international readership. NEC was always conceived as an international contract, so including UK case law would probably limit it to a UK readership.

I have included within the text many examples, as I have found when running NEC3 training courses that delegates understand principles much better if they can be given illustrated and worked examples, and where numbers are involved, using real calculations.

My personal motivation for writing a book is to collate into one volume a significant quantity of material which I have gathered over the years and to share with others a substantial knowledge and experience of the contract.

For simplicity I have used the pronouns 'he' and 'his' throughout the book; however, this does not signify any gender bias.

Acknowledgements

I would like to extend my sincere thanks and gratitude to Lesley, the love of my life and now my wife, who always gives me the love, the time, the inspiration and the support to fulfil my life's ambitions!

This book is dedicated to my granddaughter Grace. How fortunate she is to be supported in her life by the love of her parents, and also by the wonderful family and friends who have always supported me.

Introduction

0.1 Background to the NEC contracts

The history of the New Engineering Contract (NEC) stems from 1985, when the Council of the Institution of Civil Engineers (ICE) approved a recommendation from its Legal Affairs Committee 'to lead a fundamental review of alternative contract strategies for civil engineering design and construction with the objective of identifying the needs for good practice'.

There had been a concern for many years that the policy of construction contracts being sector-specific, i.e. building, civil engineering, process, and using a single procurement route was outdated. In addition, there had been concerns for many years about issues such as post-completion prolonged claims negotiations and the time taken to agree final accounts.

In 1986 the specification for the new contract was submitted to the Legal Affairs Committee, and in 1991 the first consultation document was issued, with consultation lasting a period of two years. It is perhaps worth mentioning that all of the successive NECs have gone through a consultation process, the drafters inviting comments by the NEC Users' Group, so that potential users can contribute to the proposed new contracts.

The NEC (now NEC3) is now being used substantially within the construction industry, and by many Employers in the public and private sectors in several countries. There are now several national and international flagship projects and tens of thousands of other projects which have been, and are being, built using the contract, with most Employers having reported that the contract gives far greater control of time, cost and quality issues combined with greatly improved relationships between the contracting parties.

In July 1994, Sir Michael Latham's report *Constructing the Team*[1] recommended that the then NEC should be adopted by Employers as the standard form of contract to be used in both the private and public sectors rather than using bespoke forms of contract, and suggested that it should become a national standard contract across the whole of engineering and construction work generally. He stated that endlessly refining existing conditions of contract would not solve adversarial problems, a set of basic principles is required on which modern contracts can be based and a complete family of interlocking documents is also required.

He also stated that the NEC fulfils many of these principles and requirements, but that changes to it would be desirable and the matrix was not yet complete.

Latham also stated that the most effective form of contract should include:

1. a specific duty for all parties to deal fairly with each other and in an atmosphere of mutual cooperation;
2. firm duties of teamwork, with shared financial motivation to pursue those objectives;
3. a wholly interrelated package of documents suitable for all projects and any procurement route;
4. easily comprehensible language and with guidance notes attached;
5. separation of the roles of contract administrator, project or lead manager and adjudicator;
6. a choice of allocation of risks, to be decided as appropriate to each project;
7. where variations occur, they should be priced in advance, with provision for independent adjudication if agreement cannot be reached;
8. provision for assessing interim payments by methods other than monthly valuation, i.e. milestones, activity schedules or payment schedules;
9. clearly setting out the period within which interim payments should be made, failing which there should be an automatic right to involving payment of interest;
10. providing for secure trust fund routes of payment;
11. while avoiding conflict, providing for speedy dispute resolution by an impartial adjudicator/referee/expert;
12. providing for incentives for exceptional performance;
13. making provision where appropriate for advance mobilisation payments.

The NEC family now provides for all of Latham's recommendations.

In June 2005, NEC3, the current generation of NEC contracts, was launched as a complete review and update of the whole NEC family, including the introduction of two new contracts, the Term Service Contract and the Framework Contract.

Since that date, the Supply and Supply Short Contracts and the Professional Services Short Contract have also joined the family.

Alongside the launch of NEC3, the Office of Government Commerce (OGC) announced that, after extensive research, they identified NEC3 as the contract which should be adopted by private and public Employers in commissioning their projects.

This was reflected on the inside front cover of the NEC3 contracts when they launched in June 2005, which stated:

> OGC advises public sector procurers that the form of contract used has to be selected according to the objectives of the project, aiming to satisfy the Achieving Excellence in Construction (AEC) principles. This edition of the NEC (NEC3) complies fully with the AEC principles. OGC recommends the use of NEC3 by public sector construction procurers on their construction projects.[2]

This note was changed in the 2010 reprint, and currently reads:

> The Construction Clients' Board recommends that public sector organisations use the NEC3 contracts when procuring construction. Standardising use of this comprehensive suite of contracts should help to deliver efficiencies across the public sector and promote behaviours in line with the principles of Achieving Excellence in Construction.[3]

It should be stressed that NEC3 is still the only contract to have received this recommendation to date.

0.2 Objectives of the NEC3 contracts

Whilst articles written about NEC tend to give prominence to many of the big 'flagship' projects carried out using the contracts, of much more significance are the very many more modest-size and value projects being procured, many of which have been executed using the Engineering and Construction Short Contract (ECSC), which is the subject of this book. It is also in this area that many of the problems have arisen in implementation of the contract, where perhaps less experienced practitioners may be involved; some of these problems have hopefully been addressed within the book.

The original objectives of the NEC contracts, and more specifically the ECSC, were to make improvements under three headings: (1) flexibility; (2) clarity and simplicity; and (3) stimulus to good management.

(i) Flexibility

The contract should be able to be used:

- for any engineering and/or construction work containing any or all of the traditional disciplines, such as civil engineering, building, electrical and mechanical work and also process engineering (previously, contracts had been written for use by specific sectors of the industry, e.g. ICE civil engineering contracts, Joint Contracts Tribunal (JCT) building contracts, Institution of Chemical Engineering process contracts);
- whether the Contractor has full, some or no design responsibility (previously, most contracts had provided for portions of the work to be designed by the Contractor, but with separate design and build versions if the Contractor was to design all or most of the works);
- to provide all the normal current options for types of contract, such as lump sum, remeasurement, cost-reimbursable, target and management contracts (previously, contracts were written primarily as either lump sum or remeasurement contracts, so there was no choice of procurement method when using a specific standard form);
- to allocate risks to suit each particular project (previously, contracts were

written with risks allocated by the contract drafters, and one had to be expert in contract drafting to amend the conditions to suit each specific project on which it was used);

- anywhere in the world (previously, contracts were written with country-specific clauses, which made it difficult to use them in other countries. Any country-specific NEC clauses are included as Secondary Options, e.g. Option Y(UK)2 and Y(UK)3, and in the case of the ECSC, as 'additional conditions').

To date, the NEC contracts have been used for projects as widely diverse as airports, sports stadiums, water treatment works, housing projects, in many parts of the world and even research projects in the Arctic.

(ii) Clarity and simplicity

- As in the case of the other NEC contracts, drafting has been carried out using an iterative process based on flow charts. The flow charts are published with the Guidance Notes.
- The contract is written in ordinary language and in the present tense.
- As far as possible, NEC only uses words which are in common use so that it is easily understood, particularly where the user's first language is not English. Previously obscure words such as 'whereinbeforesaid', 'hereinafter' and 'aforementioned' were commonplace in contracts! It has few sentences that contain more than 40 words and uses bullet points to subdivide longer clauses.
- Subjective and technical terms have been avoided as much as possible.
- The number of clauses and the amount of text are also less than in most other standard forms of contract, with most sentences being short, bullet points introduced to assist clarity and there is an avoidance of cross-referencing, found in more traditional standard forms.
- It is also arranged in a format which allows the user to gain familiarity with its contents, and required actions are defined precisely, thereby reducing the likelihood of disputes.
- Finally, subjective words like 'fair' and 'reasonable' have been used as little as possible as they can lead to ambiguity, so more objective words and statements are used.
- It is significant that the Plain English Commission have given their seal of approval to the ECSC, the first known occasion where this has been done for a standard form of construction contract.

Some critics of NEC have commented that the 'simple language' is actually a disadvantage as certain clauses may lack definition and there are certain recognised words that are commonly used in contracts. There is very little case law in existence with NEC contracts and, whilst adjudications are confidential and unreported, anecdotal evidence suggests that there do not appear to be any more adjudications with NEC contracts than any other, which would tend to suggest that the criticism may be unfounded.

(iii) Stimulus to good management

This is perhaps the most important objective of the NEC contracts, in that every procedure has been designed so that its implementation should contribute to, rather than detract from, the effective management of the work. In order to be effective in this respect, contracts should motivate the parties to want to manage the outcome of the contract proactively, not just to react to situations. NEC intends and requires the parties to be proactive and not reactive. It also requires the parties (and those who represent them) to have the necessary experience (sadly often lacking) and to be properly trained so that they understand how NEC works.

The philosophy is founded on two principles:

1. 'Foresight applied collaboratively mitigates problems and shrinks risk.'
2. 'Clear division of function and responsibility helps accountability and motivates people to play their part.'

Examples of foresight within the ECSC are the early warning and compensation event procedures, the early warning provision requiring the Employer and the Contractor each to notify the other upon becoming aware of any matter which could have an impact on price, time or quality.

A view held by many Employers is that an early warning is something that the Contractor would give, and is an early notice of a 'claim'. This is an erroneous view as, first, early warnings should be given by either the Employer or the Contractor, whoever becomes aware of it first, the process being designed to allow the Employer and Contractor to share knowledge of a potential issue before it becomes a problem, and, second, early warnings should be notified regardless of whose fault the problem is: it is about raising and resolving the problem, not compensating the affected party.

The compensation event procedure requires the Contractor to submit within two weeks a quotation showing the cost and any time deal effect of the event. The Employer then replies to the quotation within two weeks, enabling the matter to be properly resolved close to the time of the event rather than many months or even years later.

The programme should also be an important management document, though there is no prescribed format for the programme in the ECSC as there is with other NEC3 contracts. The Contract Data states that the Employer should state if a programme is required, what form it is to be in, what is to be shown on it and when it is to be submitted and updated.

In total the ECSC is designed to provide a modern method for Employers, Contractors, Subcontractors, Designers and others to work collaboratively and to achieve their objectives more consistently than has been possible using other traditional forms of contract. People will be motivated to play their part in collaborative management if it is in their commercial and professional interest to do so.

Uncertainty about what is to be done and the inherent risks can often lead to disputes and confrontation but the Engineering and Construction Short Contract (ECSC) clearly allocates risks and the collaborative approach will reduce those risks for all the parties so that uncertainty will not arise.

0.3 Why a short form of contract?

The majority of building and engineering contracts carried out are of relatively low risk and complexity, and often, but not necessarily, low value. Hence the work involved in preparing a full ECC cannot be justified in many cases. Also, the detailed procedures and management systems in the ECC may not be necessary in many contracts. Thus, there is a definite need for a shorter contract which is simple, easy to use and more suited to the low risk and complexity, and lower value type of contract.

The need for shorter contracts has been recognised by most of the contract writing bodies. For example, the JCT publish the *Minor Works Building Contract*, the Association for Consultancy and Engineering/Civil Engineering Contractors Association (ACE/CECA) the *ICC Infrastructure Conditions of Contract Minor Works Version* and the International Federation of Consulting Engineers (FIDIC) the *Short Form of Contract (The 'Green Book')*.

The Latham Report *Constructing the Team* emphasised the need for a shorter contract. Paragraph 5.20 of the report states that 'provision should be made for a simpler and shorter minor works document' and a footnote states: 'for example, the Landscape Institute regards the NEC as too complex even for landscape services in connection with civil engineering projects'.

The first page of the NEC3 ECSC states:

> The NEC3 Engineering and Construction Short Contract is an alternative to the Engineering and Construction Contract and is for use with contracts which do not require sophisticated management techniques, comprise straightforward work and impose only low risks on both the Employer and the Contractor.

Unlike other forms of contract, there is no limiting or recommended financial value for contracts under the ECSC, because the above criteria are not in general related to a particular sum of money. It is possible to have large-value, low-risk and simple work carried out under the ECSC. On the other hand it may not be appropriate to carry out complex high-risk work under the ECSC even though its monetary value is small. Several Employers have used the ECSC very successfully for work in excess of £2m in value.

By contrast, the many Minor Works contracts are normally recommended for work not exceeding £250,000 in value with contract periods not exceeding six months. In practice, however, these forms are often used for work in excess of the recommended values.

Intended users of the ECSC range from major organisations carrying out a large number of straightforward, low-risk projects to domestic householders seeking a simple, user-friendly contract to enable them to appoint a builder for a home extension. It has to be said that the great majority of users of the contract fall into the former category.

The history of the ECSC is as follows:

- first edition 1999;
- second edition 2005;
- reprinted 2007, 2008, 2009, 2010, 2011;
- reprinted with amendments 2013.

Since the ECSC has been drafted for use on straightforward and low-risk work, many matters covered by the ECC have been omitted. Whilst the intention has been to cater for anything which might go wrong on the simpler project, it is always possible that something may occur which is not covered in the ECSC. Such matters will then have to be dealt with by the common law of contract. A comparison with the ECC is given below.

0.4 Differences between the ECC and the ECSC

Detailed comparison with the clauses of the ECC is dealt with in the specific comments on the ECSC clauses later in this book.

However there are some important differences of a general nature. For example, the Project Manager and Supervisor are absent, but the Employer has powers to delegate as necessary to suit his organisation and his particular purposes. Also there are no main options; instead, the flexibility and allocation of risk are covered in the Price List. Neither are there any Secondary Options; the reason for this is that these would probably not be used in the type of contract for which the ECSC is designed. However, retention and delay damages, both Secondary Options in the ECC, have been included in the core clauses of the ECSC.

The number of definitions has been reduced, and since there are no main options, the Prices and the Price for Work Done to Date have standard definitions, with the flexibility being inherent in the use of the Price List. Some clauses which are unlikely to be used have been omitted but if required can be incorporated by appropriate drafting of the Works Information or the use of additional conditions of contract. For instance, no mention is made of health and safety requirements, but any specific requirements can be included in the Works Information.

There is a requirement that the Employer is to accept the Contractor's design before he can start work on the work designed, although detailed procedures for acceptance have not been included. Similarly, there is no procedure for acceptance of Subcontractors but responsibility of the Contractor for the performance of Subcontractors is clearly stated.

In the ECC there is a prescribed list of items which must be included in the Contractor's programme. All of these may not be necessary for a project using the ECSC, and therefore any detailed requirements for the programme within the ECSC are to be stated in the Works Information.

The payment procedure in the ECSC is different from that in the ECC. In the ECSC the Contractor is required to submit an application for payment, after which the Employer is obliged to pay after making any corrections he considers necessary. There is no payment certificate as in the ECC.

Also the definition of Contractor's Cost is much simplified. There is no Schedule of Cost Components; instead cost is defined in terms of payments made by the Contractor for four different items:

1.   people employed by the Contractor;
2.   Plant and Materials;
3.   work subcontracted by the Contractor;
4.   Equipment.

The list of compensation events has been reduced from 19 No in the ECC to 14 No in the ECSC. The weather event has been simplified in terms of time lost.

Assessment of compensation events is similar to that of the ECC, except that effectively the rates in the Price List are used where possible.

The section on title is much abbreviated and the title to surplus materials on the site is now with the Contractor rather than the Employer.

Section 8 on insurance etc. has been completely restructured to include the equivalent of *force majeure* events.

Reasons which give rights to termination in Section 9 have been reduced and some of the detailed procedures on termination have been omitted.

Contract strategy

The ECSC does not give the Employer the choice of contract strategy in the form that is provided in the ECC. However, the Price List can be used to produce a lump sum contract or a bill of quantities-based contract. Target contracts and cost-reimbursable contracts are not provided for as they are regarded as unsuitable and too complex for this type of work. Similarly, management contracts will not be used on the type of work for which the ECSC is designed.

When the Employer prepares the contract he must decide the strategy he wishes to employ and hence who should carry the risks of quantities and pricing.

0.5 Essential differences between the ECSC and other forms of contract

The ECSC is part of the NEC3 matrix of contracts – Engineering and Construction Contract and Subcontract, Engineering and Construction Short Contract and Short Subcontract, Professional Services and Professional Services Short

Contract, Term Service and Term Service Short Contract, Supply and Supply Short Contract, Framework Contract, and the Adjudicator's Contract – allowing all parties, whatever the project or service to be provided, to work under similar conditions.

- *Flexibility of use*: The ECSC is not sector-specific, in terms of it being a building, civil engineering, mechanical engineering or process contract: as with other NEC contracts it can be used for any form of engineering or construction. This is particularly useful where a major project or, probably more likely in the case of the ECSC, a part of a major project, such as an airport or a sports stadium, can be a combination of building, civil engineering and mechanical and electrical elements.
- *Flexibility of procurement*: The ECSC's Price List, together with flexibility in terms of Contractor design, allows it to be used for several procurement methods whether the Contractor is to design all, none or part of the works. Again, other contracts do not offer this flexibility.
- *Early warning*: The ECSC contains express provisions requiring the Contractor and the Employer to notify when either becomes aware of any matter which could affect price, time or quality. Few contracts have this express requirement.
- *Programme*: The ECSC is unlike the ECC in that there is no express requirement for a programme within the core clauses, though within the Contract Data there is a section entitled 'Requirements for the programme' which invites the contract compiler to state whether a programme is required, and if it is, what form it should take, what information is to be shown on it, when it is to be submitted and when it is to be updated. In reality, many users of the ECSC 'copy and paste' parts of the programme requirements from the ECC, including any requirement for method statements.
- *Compensation events*: This procedure is unique to NEC and requires the Contractor to price the time and cost effect of a change within two weeks and for the Employer to reply within two weeks. There is therefore a 'rolling' Final Account with early settlement and no later 'end of job' claims for delay and/or disruption. It is also more beneficial for the Contractor in terms of his cash flow as the Contractor is paid agreed sums rather than reduced 'on account' payments, which are subject to later agreement and payment.
- *Disputes*: The contract encourages better relationships and there is far less tendency for disputes because of its provisions. If a dispute should arise there are clear procedures as to how to deal with it, i.e. adjudication, tribunal.

0.6 Arrangement of the ECSC

The ECSC includes the following sections.

Contract Data

- The Contractor's Offer
- The Employer's Acceptance
- Price List
- Works Information
- Site Information.

Conditions of Contract

1. General
2. The Contractor's main responsibilities
3. Time
4. Defects
5. Payment
6. Compensation events
7. Title
8. Indemnity, insurance and liability
9. Termination and dispute resolution.

0.7 The Price List

The Price List can be used in a number of different ways:

(i)   as a lump sum contract based on a specification and/or drawings;

(ii) as a lump sum contract where the Employer gives a list of work items and the Contractor is required to price each item as a lump sum. The total of the lump sums is the lump sum offer for the works;

(iii) as an admeasurement contract in which the Employer lists the quantities and item descriptions and the tenderer enters a rate and extends the rate to the prices column. Here the itemisation is decided by the Employer and is not subject to change except by compensation events. There is no method of measurement stated. If the actual quantities carried out are different from those stated, they are corrected (see definition of Price for Work Done to Date);

(iv) as an admeasurement contract in which the Employer lists work items to be priced in the form of a lump sum, and remeasurable items as in (iii) above;

(v)   as a schedule of rates;

(vi) any of the above (i) to (v), but where the Contractor writes the Price List to the instructions of the Employer in the invitation to tender or as a result of negotiation.

(i) Lump sum contract based on specification and/or drawings

Here a brief description is given within the Price List, referenced to a specification and/or drawings, the Contractor pricing the work as a single lump sum, which can be made up of a series of individual lump sums.

Procurement using specification and drawings is still the way the largest number of projects, particularly those lower in value, are procured. In addition, many projects are procured on a design and build basis, which again can be done using this form of contract. The lump sum contract has always been the most commonly selected alternative under the NEC3 contracts.

The lump sum contract is normally used where the Employer knows exactly what he wants, and is able to clearly define it through the Works Information, which would mainly comprise drawings and specifications, but can alternatively be a performance specification, where the Contractor is to design and build the works to meet specific performance objectives.

Whilst the ECSC is appropriate for all forms of engineering and construction, the lump sum alternative is particularly appropriate for building projects where the work can more often be clearly defined.

The price may be in the form of an activity schedule, which is a list of activities, normally prepared by the Contractor, and can be used where the Employer has provided all the Works Information, or where the Contractor is to design all or part of the works based on some form of performance criteria, and therefore there is also Works Information provided by the Contractor.

Establishing and taking off the quantities of work involved to achieve the Completion of each activity would then be the responsibility of the tendering Contractors.

When the Contractor has priced the activity schedule, the lump sum for each activity can be the Price to be paid by the Employer in the assessment following Completion of that activity.

This would therefore be a stage payment contract and payment can be linked to completion of activities by writing additional conditions of contract. Administration of payment aspects is therefore fairly simple and transparent.

This alternative is ideally suited to Contractor's design but can be used for Employer's design or split design responsibility.

It is important to recognise that the activity schedule has two primary functions:

1. It shows how the tendering Contractors have built up their price and can therefore be used as part of the tender assessment process.
2. It is used to calculate the Price for Work Done to Date.

The Employer does not provide a bill of quantities with this alternative; the taking off of the quantities of work involved to achieve the completion of each activity and the pricing of the activity are the responsibility of the tendering Contractors.

There are certain advantages to the Employer using the lump sum alternative:

- There is no requirement for a bill of quantities to be prepared for tendering Contractors to price. This will save the Employer time and money at the pre-tender stage.

- The Contractor holds the risk of inaccuracies in the quantities, or missing an item of work stated within the Works Information. However, it must be reinforced that the Employer holds the risk of inaccuracies and omissions within the Works Information itself.
- The assessment of the Price for Work Done to Date is easier and quicker than with the other alternatives. There is no requirement to remeasure the work and apply quantities to rates and prices to calculate the amount due to the Contractor.
- Programme and activity schedule preparation are linked as integrated activities which would tend to lead to a more comprehensive tender.
- Payment is linked to completion of an activity or group of activities related back to the programme, so cash flow requirements for both parties are more visible.
- In order to receive payment, the Contractor has to complete an activity by the assessment date, so he has to price and programme realistically and accurately, and is motivated to keep to that programme during construction.
- The assessment of the effects of compensation events is related back to the activity schedules. Any change in resources or methods associated with an activity can be compared with those stated in the Accepted Programme before the compensation event occurred.

The disadvantages to the Employer are:

- As there is no specific document that all the tendering Contractors price, it is difficult to assess tenders on a 'like-for-like' basis. Some Employers cite that as an advantage, saying that they do not require the facility to check tenders on a line-by-line basis anyway!

  It is not uncommon, to assist the assessment of tenders, for Employers to issue templates for the Contractor to price covering the main construction elements as part of the tender, with the tendering Contractors then inserting the activities within each element. In that way, each element will have been priced on a 'like-for-like' basis, but the activities within each element may have been described and priced differently by each tenderer.

  If a template is used, Employers should be wary of the template being too prescriptive and almost being seen by tenderers as a bill of quantities.
- The tendering Contractors have to spend time, and money, in taking off the quantities of work in order to establish and populate the activity schedule. In order to reduce the tenderers' efforts in producing their own quantities, some Employers issue quantities to tenders, but these must be stressed as for indicative purposes only, and tenderers should not rely on those quantities.
- The Employer holds the risk of clearly defining the Works Information in order that the Contractor can prepare the activity schedule.

  When activity schedules are used, payment is linked to Completion of activities; the Contractor must plan and carry out his work effectively with the cash flow requirements for both parties being clearly visible. Administration of payments is therefore fairly simple.

This alternative is ideally suited to Contractor's design but can be used for Employer's design or divided design responsibility.

(ii) *Lump sum contract where the Employer gives a list of work items and the Contractor is required to price each item as a lump sum. The total of the lump sums is the lump sum offer for the works*

This is similar to pricing alternative (i), but in this case the Employer itemises the various elements of the work within the Price List referenced to a specification and/or drawings. The Contractor prices the work as a single lump sum, which can be made up of a series of individual lump sums identified by the Employer, and the Contractor prices each item.

Again, the price may be in the form of an activity schedule prepared by the Contractor or the Employer, and can be used where the Employer has provided all the Works Information, or where the Contractor is to design all or part of the works based on some form of performance criteria.

Again, establishing and taking off the quantities of work involved to achieve the Completion of each activity would then be the responsibility of the tendering Contractors.

Again, this alternative is ideally suited to Contractor's design but can be used for Employer's design or split design responsibility.

(iii) *Admeasurement contract in which the Employer lists the quantities and item descriptions and the tenderer enters a rate and extends the rate to the prices column. Here the itemisation is decided by the Employer and is not subject to change except by events. There is no method of measurement stated. If the actual quantities carried out are different from those stated, they are corrected (see definition of Price for Work Done to Date)*

This alternative is a remeasurement contract based on a bill of quantities prepared by the Employer with the Contractor pricing the items in the bill of quantities including matters which are at his risk. The Contractor is then paid for the quantity completed to date at the rates and Prices in the bill of quantities.

In the early days of the NEC, Employers would tend to select remeasurement alternatives initially as they mirrored the traditional bills of quantities-based contracts they were using at the time. Once Employers became more experienced and confident in the use of the contract, they tended to move towards other alternatives.

It is important that, if selecting a remeasurement contract, one should consider the merits of preparing and using bills of quantities against other procurement methods. An Employer should not select remeasurement just because in the past, with other forms of contract, he has always begun the procurement process by producing the drawings and specification then measuring the work. Again, this comes back to a procurement strategy consideration.

This alternative is normally used where the Employer knows what he wants and is able to define it clearly through the Works Information, and measure it within the bills of quantities, but there are likely to be changes in the quantities that may or may not be considered as compensation events.

Although there has been a general decline in use of bills of quantities in the past 25 years, the alternative is still quite widely used, particularly on civil engineering projects where the final quantity of excavations and filling materials cannot be reasonably forecast with great accuracy.

This alternative does not work well when there is a significant amount of Contractor's design. In this case, a lump sum contract should be used rather than remeasurement, as the Contractor can then carry out initial design calculations, price the work and include the various design stages within his activity schedule rather than the Employer measuring the work, which would not be practical.

A bill of quantities comprises a list of work items and quantities prepared by the Employer and priced by the Contractor. Standard methods of measurement are published (e.g. SMM7 and CESMM3 or local equivalents) which state the items to be included and how the quantities are calculated.

Confusion can sometimes arise as to whether the tendering Contractors should price the bills of quantities or the Works Information. The Contractor Provides the Works in accordance with the Works Information (Clause 20.1), therefore it is assumed that he programmes the works in accordance with the Works Information.

Information in the bill of quantities is not Works Information or Site Information; it is purely for the purposes of payment. It does not tell the Contractor what he has to do, but it must clearly be a fair representation of the Scope of Works to allow the Contractor to price the Works accurately. Although a Contractor will often use the quantities within the bill of quantities as a guide to calculate time scales, the bill is used as a basis for inviting and assessing tenders, so it is a pricing and payment document, i.e. it deals with money. It is reasonable for the Contractor to presume that the quantities are a fair representation of what is included in the Works Information.

If there is an ambiguity or inconsistency within the Works Information either Party notifies the other as soon as either becomes aware that it exists and the Employer instructs a change to correct it. This may then be addressed as a compensation event under Clause 60.1 due to the change to the Works Information, with Clause 63.8 stating that, where an instruction is given to change the Works Information to resolve an ambiguity or inconsistency, it is assessed as if the Prices and the Completion Date were for the interpretation most favourable to the Party which did not provide the Work Information.

Clearly, if the Contractor notices the error during the tender period then he should raise it as a query, then the matter can be dealt with and all tenderers informed. However, while the Contractor should notify if and when he finds an error, he is not under any obligation to look for the error!

*(iv) Admeasurement contract in which the Employer lists work items
to be priced in the form of a lump sum, and remeasurable items as in
(iii) above*

This is similar to pricing alternative (iii). Again this would create a remeasurement contract based on a bill of quantities prepared by the Employer with the Contractor pricing the items in the bill of quantities, including matters which are at his risk, but there would be lump sum items as well as measured works. Again, the Contractor is then paid for the quantity completed to date at the rates and Prices in the bill of quantities.

(v) Schedule of rates

This alternative is again a remeasurement contract based on a bill of quantities prepared by the Employer with the Contractor pricing the items in the bill of quantities, including matters which are at his risk, but for this alternative there are either no quantities or estimated quantities so that the Employer is able to compare tenders. The Contractor is then paid for the quantity completed to date at the rates and Prices in the bill of quantities.

This alternative is often used for refurbishment and repair contracts, where the quantities are not known and cannot even be estimated.

0.8 Additional conditions

All changes to the core clauses should be included as 'additional conditions' rather than amending the core clauses themselves, so in effect the clause remains in the contract, but is amended by the additional conditions. It is also critical that, when drafting an additional condition, it must be clearly stated what happens to the original core clause, for example it is deleted. If the original core clause is not deleted, then it is likely that an inconsistency will arise which will be interpreted against the party who wrote or amended the clause, i.e. the Employer.

These conditions may modify or add to the core clause, to suit any risk allocation or other special requirements of the particular contract. However, changes should be kept to a minimum, consistent with the objective of using industry standard, impartially written contracts.

It must be remembered that, if an Employer amends a contract to allocate a risk to the Contractor which may have been intended to be held by the Employer, the Contractor has a right to price it in terms of time and money. Therefore the practice of Employers amending contracts to pass risk to Contractors without considering who is best able to price, control and manage those risks can in many cases prove to be unwise and uneconomical.

It is important to spend time considering whether an additional condition is appropriate in each case. Then when that decision has been made, the clause is drafted correctly and aligned to the drafting principles of the original contract. In the case of the ECSC, use ordinary language, present tense, short sentences and bullet pointing. It is not unusual to see additional conditions in an ECSC contract written in a legalistic language, in the future tense and without punctuation apart from full stops!

Unlike the ECC, there are no Secondary Options, though it may be prudent on a project-by-project basis to consider adapting some of the Secondary Options from the ECC for use with the ECSC.

Some suggestions for additional conditions which are aligned to the Secondary Options from the ECC and which can be adapted from them include the following.

(i) Price adjustment for inflation

The default within the ECSC is that the contract is 'fixed price' in terms of inflation, i.e. the Contractor has priced the work to include any inflation he may encounter during the period of carrying out the contract and inflation is his risk.

The Employer should make the decision at the time of preparing the tender documents as to whether inflation for the duration of the contract is to be:

* the Contractor's risk – in which case the contract does not need to be amended;
* the Employer's risk – in which case, he should include some provision to adjust the Prices to take account of inflation. Changes to the Prices arising from inflation would then be dealt with as compensation events.

(ii) Changes in the law

Again, the default is that the contract is 'fixed price' in terms of changes in the law, i.e. the Contractor has priced the work to include any changes in the law he may encounter during the period of the contract. Again it is the Contractor's risk.

Again, the Employer should make the decision at the time of preparing the tender documents as to whether changes in the law for the duration of the contract are to be:

* the Contractor's risk – in which case the contract does not need to be amended;
* the Employer's risk – in which case, he should include some provision to adjust the Prices to take account of changes in the law. Changes to the Prices arising from changes in the law would then be dealt with as compensation events.

(iii) Parent company guarantee

This form of guarantee is given by a parent company (or holding company) to guarantee the proper performance of a contract by one of its subsidiaries (the Contractor), who, whilst having limited financial resources himself, may be owned by a larger and more financially sound parent company.

If a parent company guarantee is required, it should be provided by the date the contract comes into existence, or shortly afterwards.

The parent company guarantee should have an expiry date, which could be Completion of the project, the Defects Date, or may even include the six-, ten- or twelve-year limitation period following Completion of the Works to cover any liability for potential latent defects, the expiry date being defined within the Works Information. Parent company guarantees are normally used as an alternative to a performance bond.

(iv) Sectional Completion

This is a provision that many have added into the ECSC, if the Employer requires sections of the works to be completed by the Contractor before the whole of the works are completed.

References in the contract to the works, Completion and Completion Date will then apply to either the whole of the works or a section.

(v) Bonus for early Completion

Often early Completion would be of benefit to the Employer, for example the Contractor is building something within a sporting or cultural complex and early Completion would allow a sporting event or concert to take place, bringing revenue to the Employer.

Some consideration could be given for the Employer to pay a bonus to the Contractor for early Completion.

(vi) Delay damages

Delay damages in the NEC contracts are normally referred to in other contracts as liquidated damages.

Delay damages are predefined amounts which can be inserted into the contract and paid or withheld from the Contractor in the event that he fails to complete the works by the Completion Date. The amount included within the contract for delay damages should be a 'genuine pre-estimate of likely losses', not a penalty.

(vii) Performance bond

A performance bond is an arrangement whereby the performance of a contracting party (the Principal) is backed by a third party (the Surety), which could be a bank, insurance company or other financial institution, that should the Principal fail in his obligations under the contract, normally due to the Principal's insolvency, the Surety will pay a preagreed sum of money to the other contracting party (the Beneficiary).

Whilst bonds are not always linked to insolvency, contracts normally give remedies in the event of default, e.g. Defects provisions, retention, withholding of payment, delay damages. It is where the defaulter is not able to provide a remedy as he is insolvent that bonds normally provide the remedy.

This bond may be between any two contracting parties, Employer and Contractor, Contractor and Subcontractor, Contractor and Supplier, or any other contractual relationship.

If a performance bond is required from the Contractor it should be provided by the Contract Date. The amount of the bond, usually 10 per cent of the contract value, must be stated and the form of the bond must be in the form stated in the Works Information. In the event of the default by the Contractor the Employer is not paid the full amount of the bond, but the amount of the bond is the maximum amount available to meet the Employer's costs incurred as a direct result of the default.

(viii) Limitation of the Contractor's liability for his design to reasonable skill and care

Whilst the Works Information defines what, if any, design is to be carried out by the Contractor, the contract is silent on the standard of care to be exercised by the Contractor when carrying out any design.

Two terms that relate to design liability are 'fitness for purpose' and 'reasonable skill and care'.

In construction, fitness for purpose means producing a finished project fit in all respects for its intended purpose. This is an absolute duty independent of negligence, and in the absence of any express terms within the contract to the contrary, a Contractor who has a design responsibility will be required to design and build the project 'fit for purpose'.

'Fitness for purpose', by default, is the standard of care required if the Contractor is to design any element of the works, but the Employer may consider 'relaxing' this to 'reasonable skill and care'. (See also Chapter 2.)

(ix) Key Performance Indicators

In recent years, Key Performance Indicators (KPIs) have become a regular provision within construction contracts. The performance of the Contractor is monitored and measured against KPIs, with incentives set out in the contract for exceptional performance.

The Contractor normally reports his performance against the KPIs to the Employer at intervals stated in the contract and is paid the amount stated in the contract if the target for a KPI is improved upon or achieved. Note that there is no payment due from the Contractor if he fails to achieve a stated target.

Normally, the Employer may add a new KPI and associated payment to the contract, but may not delete or reduce a payment.

Typical KPIs

• Improved client satisfaction
• Employee satisfaction

- Reduction of defects
- Savings in construction time
- Savings in construction cost
- Improved productivity
- Improved profitability
- Reduced health and safety incidents
- Reduced staff turnover
- Sickness absence
- Improved qualifications and skills
- Reduced impact on environment
- Whole life performance
- Reduced waste.

It is critical in setting KPIs that the SMART principle applies:

- *Specific* – It has to be clear what each KPI actually measures. A KPI to measure how effectively people generally do their job will not be effective. It must cover a certain aspect of what they do, e.g. specific deliverables.
- *Measurable* – Each KPI has to be measurable so that one can compare actual values with original baseline values.
- *Achievable* – Each KPI has to be achievable; one cannot aim for a goal that no one will ever achieve.
- *Relevant* – Each KPI must be relevant to the task in hand in order that improvement in delivering that task can be monitored and measured.
- *Time-based* – Each KPI must be time-based, ideally measured on a periodic basis.

(x) The Housing Grants, Construction and Regeneration (HGCR) Act 1996 and The Local Democracy, Economic Development and Construction (LDEDC) Act 2009

This option would only be applicable to UK contracts where the HGCR Act 1996 applies.

Whilst this book avoids discussing NEC3 specifically in connection with UK construction law, it is worth mentioning this Act and the LDEDC Act 2009, as they can have a direct bearing on the contract.

The LDEDC Act, specifically Part 8, amends Part II of the HGCR Act 1996, and came into force in England and Wales on 1 October 2011 and in Scotland on 1 November 2011.

It is important to note that, as the new Act amends the HGCR Act 1996, you have to take into account both Acts to understand fully how it applies to your contract.

The Act applies to all construction contracts, including:

- construction, alteration, repair, maintenance, etc.;
- all normal building and civil engineering work, including elements such as temporary works, scaffolding, site clearance, painting and decorating;

- consultants' agreements concerning construction operations;
- labour-only contracts;
- contracts of any value.

It excludes:

- extraction of oil, gas or minerals;
- supply and fix of Plant in process industries, e.g. nuclear processing, power generation, water or effluent treatment;
- contracts with residential occupiers;
- header agreements in connection with private finance initiative contracts;
- finance agreements.

The HGCR Act 1996 only applied to contracts in writing, but the LDEDC Act 2009 now also applies to oral contracts.

Briefly, the changes introduced under the LDEDC Act include:

- A notice is to be issued by the payer to the payee within five days of the payment due date, or by the payee within not later than five days after the payment due date; the amount stated is the notified sum. The absence of a notice means that the payee's application for payment serves as the notice of payment due and the payer will be obliged to pay that amount.
- A notice can be issued at a prescribed period before the final date for payment which reduces the notified sum.
- The Contractor has the right to suspend all or part of the Works, and can claim for reasonable costs in respect of costs and expenses incurred as a result of this suspension as a compensation event.
- Terms in contracts such as 'the fees and expenses of the Adjudicator as well as the reasonable expenses of the other party shall be the responsibility of the party making the reference to the Adjudicator' have been prohibited. Much has been written about the effectiveness of such clauses, and whether they comply with the previous Act, so the new Act should provide clarity for the future.
- The Adjudicator is permitted to correct his decision so as to remove a clerical or typographical error arising by accident or omission. Previously he could not make this correction.

N.B.: The publishers have issued a brief amendment to the ECSC to align it with the LDEDC Act 2009.

(xi) Option Y(UK)3: Contracts (Rights of Third Parties) Act 1999

This option would only be applicable to UK contracts where the Contracts (Rights of Third Parties) Act 1999 applies.

The principle of privity of contract is that a person who is not a party to a contract cannot enforce the terms of that contract.

The Contracts (Rights of Third Parties) Act 1999 effectively abolishes privity of contract as an absolute principle, enabling a person who is not a party to a contract, a 'third party', to enforce a term of the contract if the contract expressly provides that he may. The third party must be expressly identified in the contract, either by name or as a member of a class or by a particular description, but does not need to be in existence at the date of the contract.

Under the Act, there is available to the third party any remedy that would have been available to him in an action for breach of contract as if he had been a party to the contract.

The parties to the contract are prohibited from making any agreement to vary the contract in such a way as to adversely affect the third party's entitlement to enforce the contract without the third party's consent.

0.9 The roles of the parties

There are a number of parties referred to in the ECSC:

1. the Employer (named in the contract);
2. the Contractor (named in the contract);
3. the Employer's Representative (not named in the contract);
4. the Designer(s) (not named in the contract);
5. the Adjudicator (may be named in the contract).

The Employer

The Employer is the Client, the Parties to the contract being the Employer and the Contractor.

The Employer is rarely mentioned in most construction contracts, as he is represented by the Project Manager (ECC), Engineer, Architect or some other party, but in the ECSC the Employer represents himself, though he will usually delegate to someone to represent him (see Employer's Representative, below).

Actions which cannot be delegated by the Employer are:

- allowing access to and use of the site, as the site belongs to the Employer;
- making payment to the Contractor, as the Employer is the other Party to the contract.

The Contractor

The Contractor and the Employer are the parties to the contract. The Contractor's responsibility is to Provide the Works in accordance with the contract, the Works Information describing the works and any constraints on how he has to provide.

There are also obligations in respect of complying with instructions, notifying early warnings, submitting programmes for acceptance and notifying and submitting quotations for compensation events, which are highlighted in later chapters.

The Employer's Representative

Unlike the NEC3 ECC, there is no 'Project Manager' role, or for that matter any 'Supervisor' role under the ECSC, all actions being directly between the Employer and the Contractor. However, under Clause 14.4 the Employer, after notifying the Contractor, may delegate any actions and may cancel any delegation.

Though the term 'Employer's Representative' is not stated anywhere in the contract, someone may be appointed by the Employer, with the authority of the Employer delegated to him to manage the contract on his behalf.

Other forms of contract tend to name an Engineer or Architect who acts for the Employer, and in addition to having design responsibilities, administers the contract. The ECSC, like the other NEC3 contracts, separates the role of Designer from that of managing the project and administering the contract.

An Employer's Representative may be appointed from the Employer's own staff or may be an external consultant. If a consultant, he may be appointed using the NEC3 Professional Services Contract, though this is not mandatory. The Employer's Representative is a named individual, not a company.

It is vital that the Employer appoints a representative who has the necessary knowledge, skills and experience to carry out the role, which includes acceptance of designs and programmes, issuing early warnings, certifying payments and dealing with compensation events. Disciplines which have carried out the role to date include Engineers from a civil engineering, structural or process background, Architects, Building Surveyors and Quantity Surveyors, in addition to Project Managers themselves.

It is also vital that the Employer gives any representative full authority to act for him, particularly when considering the time scales imposed by the contract. If he has to seek approvals and consents from the Employer, then he must do so, and the appropriate authority must be given, in compliance with the contractual time scales. If the approval process is likely to be lengthy, it is critical that both the Employer and his representative set up an accelerated process to comply with the contract or that the time scales in the contract are amended. The former is the vastly preferred method in order that NEC3 will work to its full effect.

Although the Employer's Representative manages the contract at the post-contract stage, he is usually appointed pre-contract to deal with matters such as feasibility issues, advising on design, procurement, cost-planning tendering and programme matters. Note that, unlike many other contracts, the NEC3 contracts do not name the Quantity Surveyor, so this would need to be considered in dealing with financial matters pre- and post-contract. Either the Employer or his representative may carry out the role of the Quantity Surveyor himself or delegate to another.

As with any other notification under the contract, if the Employer wishes to delegate, the notice to the Contractor must be in writing, i.e. by letter or email, not by verbal communication such as a telephone call.

The delegation may be due to the Employer being absent for a period due to holidays or illness, or because the Employer wishes to appoint someone to assist him in his duties.

Although the contract is not specific, the notice should identify who the Employer is delegating to, what their authority is, and how long the delegation will last, so the Contractor is in no doubt as to who has authority under the contract.

It is important to recognise that, in delegating, the Employer is sharing an authority with the delegate who will then represent him; he is not passing responsibility on to the delegate. The authority is shared, but ultimately responsibility will remain with the Employer as the principal.

Note that there is no provision for the person with delegated authority to delegate further to others; the only delegation is by the Employer.

It is also important to note that one can only delegate outwards from the principal, i.e. no one can assume authority possibly because, within an organisation, that person is senior to the Employer.

If the Employer wishes to replace the representative he must first notify the Contractor of the name of the replacement before doing so.

Designer(s)

Designers are not named parties in the ECSC as the design may be carried out on behalf of the Employer and/or the Contractor. The Employer is required to appoint Designers for his own design, and the Contractor for his design. The concept of novation of Designers is considered in Chapter 2.

It is important to recognise that the Designer does not have the same authority as, say, an Architect under the JCT contracts: the Architect has the dual role of Designer and Contract Administrator. Under the ECSC the Designer only designs; he does not administer the contract, and therefore any issuing of drawings, other design information and associated instructions must be done through the Employer.

Again, the NEC3 Professional Services Contract may be used for appointment of Designers, though its use is not mandatory.

The Adjudicator

The Adjudicator can be chosen by one of three methods:

1. By the Employer specifically naming the individual in the Contract Data. The Contractor therefore knows at the time of tender who the Employer has named and can therefore confirm if he does not agree with the Employer's choice. This has advantages in that the parties always know who will be the

Adjudicator if they have a dispute. The disadvantages are first, that parties have commented in the past that naming the individual within the Contract Data assumes that there will be a dispute, and often the Adjudicator will require payment for having his name in the contract and being available should the parties have that dispute!

2.   By the Employer naming an Adjudicator Nominating Body in the Contract Data. This would normally be a professional institution. This tends to be the favoured method.

3.   By the parties agreeing who will be the Adjudicator at the time of the dispute. The disadvantage of this method is that once the parties are in dispute they probably would not readily agree with each other on anything!

The Adjudicator becomes involved only when either contracting party refers a dispute to him. As an impartial person he is required to give a decision on the dispute within stated time limits. If either party does not accept his decision they may refer the dispute to the tribunal (litigation or arbitration). The Adjudicator's Contract requires that payment of the Adjudicator's fees is shared by the parties unless otherwise stated. (See also Chapter 10.)

0.10 Mutual trust and cooperation

The first clause of all the NEC3 contracts requires the parties to act in a spirit of mutual trust and cooperation, the ECSC requiring the Employer and the Contractor to do so.

This mirrors Sir Michael Latham in his report *Constructing the Team*, when he recommended that the most effective form of contract should include 'a specific duty for all parties to deal fairly with each other and in an atmosphere of mutual cooperation'.

It is unusual in standard contracts in that it essentially specifies an attitude to be adopted by the parties as an obligation. It is worth considering this obligation in more detail, as the clause has often been viewed with some confusion, and for those who have spent many years in the construction industry, with a great degree of scepticism.

The first part, 'the Employer and the Contractor shall act as stated in the contract' is straightforward, and many would say is not required to be expressly stated as the parties have certain obligations within the contract, but the second part 'and in a spirit of mutual trust and cooperation' has caused some debate.

Most practitioners state that their understanding of the second part of this clause is that the parties should be non-adversarial towards each other, acting in a collaborative way and working for each other rather than against each other, and in reality that is what the clause requires, and in reality is how all parties to all contracts should behave.

However, the difficulty is that, if a party does not act in a spirit of mutual trust and cooperation, what can another party who may be affected do? The answer is that the clause is almost unenforceable as it is virtually impossible to define and quantify the breach, or the ensuing damages that flow from the breach.

To that end, Employers have been seen to insert an amendment in the ECSC to delete Clause 10.1; however, it is strongly recommended that it should remain in the contract, if merely viewed as a statement of good intent. Cynics may say if you delete the clause the parties are not required to act in a spirit of mutual trust and cooperation, but that view can only remain in the domain of cynics!

In effect, a clause requiring parties to act in a certain spirit will probably not, on its own, have any real effect. Within the ECSC, it is the clauses that follow within the contract which require early warnings, clearly detailed programmes which are submitted for acceptance and a structured change management process, that actually create and develop that level of mutual trust and cooperation rather than simply inserting a statement within the contract requiring the parties to do so.

0.11 Some unique NEC3 terms

There are some terms which, whilst not being used in other contracts, have different meanings within NEC3 contracts.

(i) Equipment and Plant

The ECC definitions of Equipment and Plant are different to many other contracts.

Equipment: Clause 11.2(6)

Equipment is defined in the contract as 'items provided by the Contractor, used by him to Provide the Works and not included in the works'.

In that sense it includes anything which is used to build the project, but which is not incorporated into the works and does not remain on Site after Completion. Specific examples of Equipment would include excavators, dumpers, lorries, cranes, tools, scaffolding and temporary accommodation. These items are normally referred to as 'Plant' in other contracts.

It also includes associated items such as fuel, lubricants, tyres and other consumables.

Plant: Clause 11.2(8)

Plant (and Materials) are 'items intended to be included in the works', specific examples being mechanical and electrical installations.

It is critical that practitioners recognise the difference between Equipment and Plant and continue to use the correct terms as they are defined separately throughout the contract.

(ii) Others

The ECC uses the term 'Others' to define parties who are 'not the Employer, the Project Manager, the Supervisor, the Adjudicator, the Contractor or any Employee, Subcontractor or Supplier of the Contractor'. Essentially, 'Others' are anyone outside the contract; for example, other contractors, statutory bodies, regulatory authorities, utilities companies.

References to Others include the Contractor's obligation to cooperate with Others in obtaining and providing information, cooperating with Others, obtaining approval of his design from Others where necessary, the order and timing of the work of Others, loss or damage to Plant and Materials supplied by Others.

Whilst the ECSC does not include or define 'Others', clearly the Contractor has an obligation to engage with and coordinate with Others as required by the contract.

(iii) Communications: Clause 13

The NEC3 contracts have strict rules regarding communications under the contract. All communications under the contract are required to be in writing and will have effect when received at the address notified by the recipient for receiving communications.

This is slightly more restrictive than the ECC, which requires that communications must be in a form which can be 'read, copied and recorded', which implies acceptance of electronic means, for example email, or via a project intranet, rather than 'in writing', which implies a more traditional hard-copy approach.

This requirement is particularly important in respect of instructions as the contract does not recognise verbal instructions and whereas under other forms of contract a verbal instruction can be confirmed by the Contractor and, if not dissented from, normally within 7 or 14 days it becomes an instruction, the NEC3 contracts do not have that provision. The instruction may be in the form of a letter or a pro forma instruction.

The Employer or the Contractor replies within the period for reply stated in the Contract Data, unless the contract states otherwise.

If the Employer is required to accept or not accept something, he should state his reasons for non-acceptance. This is particularly relevant in respect of acceptance of Contractor's design (Clause 20.2).

Note that withholding acceptance for a reason not stated in the contract is not a compensation event under the ECSC, but it is under the ECC!

It is important that notifications under the contract should be communicated separately from other communications, so for example, an early warning notice should not be included as part of a set of meeting minutes, or a letter which includes other subjects.

0.12 Subcontracting

The ECSC has been designed on the assumption that work may be subcontracted with Clause 21 dealing specifically with the appointment of Subcontractors. Unlike the ECC, the ECSC does not require 'acceptance' of Subcontractors. But Clause 21.1 makes clear that the Contractor is responsible for the work of Sub-contractors regardless of whether he sublet work.

Also the contract applies as if a Subcontractor's employees and equipment were the Contractor's own, again emphasising that the Contractor is wholly liable to the Employer for any Subcontractors.

As with other NEC contracts, there is no provision for nominated subcontracts.

The NEC3 Engineering and Construction Short Subcontract (ECSS)

Where the subcontract works comprise straightforward work, and impose low risks on the Contractor and the Subcontractor, the ECSS may be used. Where the risks are higher and/or the work is not straightforward then the Contractor may use the Engineering and Construction Subcontract (ECS), but this tends to be an exception.

The NEC3 ECSS is almost identical to the ECSC, providing back-to-back provisions, but includes a number of changes appropriate to subcontracts.

The Short Subcontract can be used by a Contractor appointed under the ECSC or the ECC. If the latter, there is provision for naming the Project Manager and Supervisor under the main contract (ECC) if this is the case, though it must be stressed that the Project Manager and Supervisor under the main contract have no authority over the Contractor or the Subcontractor under the subcontract.

Under the Short Subcontract, the following differences apply.

Subcontract Data

- 'Subcontract Data' instead of 'Contract Data';
- 'Subcontract works' instead of 'Contract works';
- Period for reply:
    - for a reply by the Contractor 'x' weeks more than the period for reply in the main contract;
    - for a reply by the Subcontractor 'x' weeks less than the period for reply in the main contract;

- References to:
    - 'if the defects date is calculated from completion of the subcontract works';
    - 'if the defects date is calculated from completion of the works in the main contract';

- References to:
    - the adjudicator;
    - the main contract adjudicator;
- The period within which the Contractor replies after the Subcontractor's submission of a quotation is:
    - Seven weeks if the main contract is under the ECC;
    - Four weeks if the main contract is under the ECSC.

General

- The Parties are the Contractor and the Subcontractor.

The Subcontractor's main responsibilities

- Clause 21 related to 'Subcontracting' now refers to 'Subsubcontracting'.

Payment

- Each payment is made within three weeks of each assessment date.

Compensation events

- The compensation events are the same excepting that the reference to the Employer is replaced by the Contractor.
- The Subcontractor submits quotations within one week of being instructed to do so by the Contractor. The Contractor replies within the period stated in the Subcontract Data, i.e.:
    - Seven weeks if the main contract is under the ECC;
    - Four weeks if the main contract is under the ECSC.

Nominated Subcontractors

Although many other contracts provide for Nominated Subcontractors chosen by the Employer, or his representative, and the Contractor is instructed to use them, the NEC contracts have never done so.

This is for a number of reasons:

- The Contractor should be responsible for managing all that he has contracted to do. Nomination will often split responsibilities.
- The use of Prime Cost Sums in bills of quantities means that the Contractor does not have to price that element of the work other than for profit and

attendances. The more Prime Cost Sums, the less the tenderers have to price and thus there is less pricing competition between the tenderers.

- Contracts will often give the Contractor relief in the form of an extension of time in the event of a default by the Nominated Subcontractor, provided that Contractor has done all that would be reasonable or practicable to manage the default.
- Contracts will often give the Contractor relief in the form of loss and expense where it cannot be recovered from the Nominated Subcontractor, again provided the Contractor had done all that would be reasonable or practicable to manage the default. This would include the insolvency of a Nominated Subcontractor and a subsequent renomination.

0.13 Appendices to Introduction

Schedule of Employer actions

The Employer

GENERAL

10.1 acts as stated in the contract and in a spirit of mutual trust and cooperation;
13.2 replies to a communication within the period for reply;
14.2 may give an instruction which changes the Works Information;
14.4 may delegate any of his actions, after notifying the Contractor;
15.1 allows access to and use of the site as necessary;
15.2 provides services and other things as stated in the Works Information;
16.1 gives an early warning by notifying the Contractor;
16.2 cooperates with the Contractor in making and considering proposals.

The Contractor's main responsibilities

20.2 accepts or does not accept the Contractor's design;
21.3 may instruct the Contractor to remove an Employee.

Time

30.3 decides the date of Completion;
30.4 may instruct the Contractor to stop or not to start any work;
31.1 receives the Contractor's programme as stated in the Works Information.

Testing and Defects

40.1 may instruct the Contractor to search for a Defect;

40.2 may notify a Defect to the Contractor;

41.4 issues the Defects Certificate to the Contractor;

42.1 assesses the cost of having a Defect corrected by other people.

Payment

50.1 receives the Contractor's application for payment;

50.4 corrects any wrongly assessed amount due;

51.1 makes a payment three weeks after the next assessment day.

Compensation events

61.1 may notify the Contractor of a compensation event;

61.1 is notified by the Contractor of a compensation event;

61.2 notifies the Contractor of his decision that the Prices and the Completion Date are not to be changed;

61.2 instructs the Contractor to submit a quotation for a compensation event;

61.3 notifies the Contractor that he did not give an early warning;

61.4 states assumptions about an event too uncertain to be forecast reasonably, and notifies a correction to an assumption;

62.2 may instruct the Contractor to submit a quotation for a proposed instruction;

62.3 assesses a compensation event if the Contractor does not provide a quotation;

62.4 replies within two weeks of receipt of a Contractor's quotation;

62.5 assesses a compensation event if he does not agree with a revised quotation or none is received;

62.6 may instruct the Contractor to submit alternative quotations;

63.5 assesses a compensation event as if the Contractor had given early warning.

Title

70.1 instructs the Contractor in respect of objects of value or historical interest.

Termination

90.1 notifies or is notified of a reason for termination and issues Termination Certificate;

90.2 may terminate (or the Contractor may terminate) (Reason 1);

90.3 may terminate after first notifying the Contractor that he has defaulted (Reason 2, 3 or 4);

90.4 may terminate for any other reason (Reason 5);

90.5 may terminate (Reason 8).

Notes

1 M. Latham (1994). *Constructing the Team: Final Report of the Government Industry Review of Procurement and Contractual Arrangements in the UK Construction Industry*. HMSO, London.
2 NEC3 *Engineering and Construction Contract* (2005). Thomas Telford, London.
3 NEC3 *Engineering and Construction Contract* (2010). Thomas Telford, London.

1 Early warnings

1.1 Introduction

Early warnings are covered within the Engineering and Construction Short Contract (ECSC) by Clauses 16.1 and 16.2.

Early warnings are an essential and valuable risk management tool within the NEC3 contracts. The ECSC obliges the Contractor and the Employer to notify each other as soon as either becomes aware of any matter which could affect the project in terms of time, cost or quality, bringing into the open as early as possible any matter which could adversely affect the outcome of the contract.

Under Clause 16.1 the Contractor and the Employer give an early warning by notifying the other as soon as either becomes aware of any matter which could:

- *increase the total of the Prices.*
  Increase the Price of the works.
- *delay Completion.*
  Delay Completion of the whole of the works.
- *impair the performance of the works in use.*
  Have a qualitative effect on the finished project.

This sometimes causes confusion, but let us look at an example. On an Employer-designed project, the Contractor is instructed to install a particular type of pump. If the Contractor knows from experience that that pump would probably not be sufficient to meet the Employer's requirements once the works are taken over, then the Contractor should give an early warning stating his concern.

The Contractor may also give an early warning by notifying the Employer of any other matter which could increase his total cost.

One could query whether a matter which could increase the Contractor's cost, but not affect the Price, should be an early warning matter, or for that matter, whether it should be anything to do with the Employer, particularly if the Contractor is not looking to recover this additional cost through the contract, but the words are 'the Contractor may give an early warning', so he is not obliged to do so.

Under Clause 16.2, after an early warning has been given, the Parties have an obligation to discuss the problem and decide how best to resolve, avoid or reduce it.

The Parties are expressly not required to give an early warning for which a compensation event has previously been notified.

The early warning procedure obliges people to be 'proactive', dealing with risks as soon as the Parties become aware of them, rather than 'reactive', waiting to see what effect they have, then trying to deal with them when it is often too late. Encouraging the early identification of problems by both Parties puts the emphasis on joint solution finding rather than blame assignment and contractual entitlement.

It is one of the most important and valuable aspects of the contract and it is perhaps surprising that, whilst some other contracts refer to an early warning process, only the NEC contracts set out in clear detail what the parties are obliged to do, with appropriate sanctions should the parties not comply (Clauses 61.3 and 63.5).

1.2 Notifying early warnings

The contract requires (Clause 13.1) that all communications under the contract must be in writing, so early warnings should not be a verbal communication such as a telephone conversation. If the first notification is a telephone conversation, or a comment in a site meeting, it should be immediately confirmed in writing to give it contractual significance (Figure 1.1).

There are some key words within the obligation to notify:

- 'The Contractor and the Employer' – no one else has the authority or obligation to give an early warning. The Employer is therefore notifying on behalf of himself, and anyone who represents him within the contract, e.g. Designers. The Contractor is notifying on behalf of himself, his Subcontractors, his Designers (if appropriate), and again many possible others whom he represents under the contract. Early warnings should be notified by the key people acting for the Parties.

  Employers are often criticised for seeing early warnings as something the Contractor has to do, and in fact most early warnings are actually issued by the Contractor. However, the Contractor and the Employer are obliged to give early warnings each to the other, so it is critical that Employers play their part in the process.

  As an example, if the Employer becomes aware that he will be late in delivering some design information to the Contractor, he should issue the early warning as soon as he becomes aware that the information will not be delivered to the Contractor, not wait and subsequently blame the Contractor for not giving an early warning stating that he has not received the information!

Contract:	EARLY WARNING NOTICE
Contract No:	EWN No..

Section A: Enquiry

To: Employer/Contractor

Description

This matter could:

☐   Increase the total of the Prices

☐   Delay Completion

☐   Impair the performance of the works in use

Date: ....

Signed:................................(Contractor/Employer)   Date: ....

Action by: Date required:

Section B: Reply

To: Contractor/Employer

Signed:................................ (Contractor/Employer)   Date: ....

Copied to:

Contractor   Employer   File   Other ....

Figure 1.1 A typical example of an early warning notice template.

- 'As soon as' – means immediately. There are a number of clauses within the contract that deal with the situation where the Contractor did not give an early warning. Whilst the party who gives the early warning must do so as soon as he becomes aware of the potential risk, the other party should respond as soon as possible and in all cases within the period for reply in Contract Data Part 1.

- 'Could' – not 'must', 'will' or 'shall'. Clearly there is an obligation to notify even if it is only felt something may affect the contract, but there is no clear evidence that it will.

The Employer enters early warning matters in the Risk Register. As stated previously, early warning of a matter for which a compensation event has previously been notified is not required.

It must be emphasised that early warnings are not the first step towards a compensation event, as is often believed. Early warnings feature in a completely separate section of the contract and, in fact, the early warning provision is intended to prevent a compensation event occurring or at least to lessen its effect. It can also be used to notify a problem which is totally the risk of the notifier. It is also worth mentioning that early warnings are a notice of a future risk, not a past one.

The Parties are not required, nor is it of any value, to notify a risk that has already happened.

Box 1.1 Example

One week before the Completion Date on a small refurbishment project, the Contractor is informed by his electrical Subcontractor that they have insufficient specialist light fittings to complete the works and that this could potentially cause a delay in Completion.

Should the Contractor issue an early warning to the Employer, bearing in mind that the failure is on the part of the Subcontractor, and probably the Contractor, and the Employer holds no liability for the lack of materials?

Yes, he should issue an early warning as it could delay Completion.

Of course, there is no entitlement on the part of the Subcontractor or the Contractor in terms of a compensation event, but the Subcontractor should issue an early warning to the Contractor, and in turn the Contractor to the Employer.

Note that, under Clause 16.2, the parties should cooperate in making and considering proposals as to how the matter can be avoided or reduced. For example, can the Subcontractor install an alternative, possibly on a temporary basis, to allow Completion to take place?

1.3 Remedy for failure to give early warning

Under Clauses 61.3 and 63.5 related to compensation events, if the Employer decides that the Contractor did not give an early warning of the event which an experienced Contractor could have given, he notifies this to the Contractor when he instructs him to submit a quotation. If the Employer has done so, the event is assessed as if the Contractor had given early warning.

Box 1.2 Example

On 1 February, the Contractor becomes aware of an issue which could increase the total of the Prices and delay Completion. He should immediately give an early warning to the Employer. However, he fails to do so until 22 February, three weeks later.

When the Contractor gives the early warning, the Employer notifies a compensation event and instructs the Contractor to submit a quotation, at the same time notifying the Contractor that he did not give an early warning of the matter that an experienced Contractor could have given.

If the Employer has given such notification, then the compensation event is assessed as if the Contractor had given early warning, so the Contractor should price his quotation based on the event as at 1 February, rather than at 22 February.

1.4 Dealing with risk

Even on a straightforward, low-risk project such as that envisaged for the ECSC, the parties must be fully aware of the principles of risk.

Risks can be classified in the ECSC as:

- risks which are owned by the Employer and do not affect the Contractor;
- risks which are owned by the Employer and do affect the Contractor in terms of Price and/or Completion Date, and are therefore included and managed as compensation events;
- risks which are owned by the Contractor and are therefore deemed to have been included within his Price and programme; these are risks which may affect the Contractor but are not listed as compensation events.

General principles of risk management

There are a number of definitions of risk, an example being:

Any uncertain event or set of events that, should it or they occur, would have an effect on achieving a party's objectives, in terms of presenting a negative threat or preventing a positive opportunity.

The objectives in a construction project are to build safely and to the required quality, to complete within the required time scale, and within the required budget.

Risk is always inherent in any activity, no matter how simple or complex it may be. However, the degree of risk will vary depending upon the circumstances involved and the amount of risk that one is willing to assume is normally directly proportional to the amount of reward that is anticipated. The greater the anticipated reward, the greater the risk one is normally willing to assume.

Risk is an integral part of the construction process and an important factor in the determination of construction costs and durations.

The parties to a construction contract assume that the contract will not only provide them with adequate protection, but also provide a fair and workable framework within which a vastly complex project can be constructed to the mutual satisfaction of all concerned.

The construction contract itself should be clear, concise, complete, legally correct and equitable. It should address the risks directly and attempt to eliminate as much of the ambiguity as possible.

Characteristics causing construction projects to present risks

All construction projects can be regarded as having risk as an underlying factor, though there are a number of differences between a construction contract and a manufacturing process, in that every construction project:

- is different, the only similarities being the people who manage them and the materials and equipment used;
- is not produced in a factory environment, but on a Site which is subject to fairly predictable seasonal variations, but unpredictable weather exceptions;
- brings a temporary multi-skilled and experienced team together, brought together for the purpose of constructing the project and possibly never working together again.

The following is a thought-provoking quotation which, while written some time ago and in a 'comical' style, still holds a certain degree of truth:

> As teams go the so-called project team really is rather peculiar, not at all like a cricket eleven, more like a scratch bunch consisting of one batsman, one goal keeper, a pole vaulter and a polo player. Normally brought together for a single enterprise, each member has different objectives, training and techniques and different rules. The relationship is unstable, even unreliable, with very little functional cohesion and no loyalty to a common end beyond that of each member coming through unscathed.[1]

As a result of the above, one can easily draw the following conclusions:

- It is impossible to produce a design which will not require subsequent modifications, either because of Client changes or because the design is found in practice not to work.

- It is impossible for a Contractor to produce a tender which will correspond exactly to the cost plus overheads and profit which will actually be incurred during the construction.
- It is impossible for a Contractor to produce a programme which can in all respects apply without revision and updating.
- When people are faced with such uncertainty they may, in an effort to defend their own beliefs and organisations, engage in conflict situations.

As a consequence, it is clear that both Employer and Contractor organisations are faced with risks that occur as a result of the nature of construction.

From the moment the decision to begin to design is taken until the new facility is in use, the Employer and Contractor take on board risks.

In addition to inherent risks within the construction risks there are risks that will arise from the agreement between the parties. These contractual risks are primarily expressed within the contract as intended between the parties, but they can also arise through inexperienced drafting which can lead to lack of contract clarity, ambiguities and inconsistencies.

The main principles of apportioning risk in construction contracts are as follows:

- The Employer will always pay for risk, either by retaining the risk within the contract and paying for it should it occur, or the Contractor prices it within his tender.
- Risks should be apportioned to the party best able to manage and/or control them and to sustain the consequences if they materialise.
- Risks outside the Contractor's control and/or foreseeability should remain with the Employer.
- The party carrying the risk should be motivated to manage the risk, possibly through a risk/reward process.

The major types of risks that can arise on construction projects are as follows.

Risks to Employer

TECHNICAL

- incomplete or incorrect design;
- inadequate site investigations.

LOGISTICAL

- availability of resources;
- availability of transportation.

RESOURCES

- insufficient resources and expertise;
- weather and seasonal problems;
- industrial relations problems.

TIMING

- inability to give access or possession.

FINANCIAL/CONTRACTUAL

- inflation;
- delay in payment;
- non-availability of funds;
- cash flow problems;
- ambiguities and/or inconsistencies in contract documents;
- work carried out by directly employed parties;
- losses due to default of Contractors.

SAFETY/RELIABILITY

- delays due to having to satisfy requirements.

QUALITY ASSURANCE/STANDARDS

- quality issues.

CHOICE OF SITE/PHYSICAL

- ground conditions;
- ground water conditions;
- exceptional weather conditions;
- damage to neighbouring buildings;
- soil strata type and variability;
- contamination.

ENVIRONMENTAL AND PLANNING

- existing rights of way;
- public enquiries;
- planning requirements;
- noise abatement;
- availability of services.

POLITICAL

- changes in Government/new legislation;
- customs and import restrictions.

DESIGN

- suitability and adaptability of the design.

UNFAMILIARITY

- pioneer design;
- experience of design team.

INFLATION

- differential inflation due to market situation at time of tendering;
- abnormal inflation.

Errors in tender documents

CONSTRUCTION

- bankruptcy/insolvency;
- industrial action.

VARIATIONS

- effect on construction duration and Price of late variations;
- effect of number of orders.

Construction delays

- delay by specialist Contractors and statutory authorities;
- delay in giving instructions to the Contractor.

Force majeure

- fundamental risk, which cannot be priced.

The risk management process

Risk management is a structured approach to identifying, assessing and managing risk.

It can be defined as:

The systematic application of management policies, procedures and practices to the tasks of identifying, analysing, assessing, treating and monitoring risks.

The risk management process can help a party analyse the degree of uncertainty in achieving the contract/project objectives, understand the consequences of particular events occurring and develop management actions to control the event or minimise its consequences.

There are six primary activities in the process.

(i) Risk identification

Risks should be identified as early as possible and continuously monitored to highlight new risks and the passing of old risks. Risk identification should initially be carried out by analysing all available documentation and knowledge as a brainstorming session involving a number of individuals, ideally chaired by an independent person external to the project. This will ensure maximum and unbiased contributions from all parties.

(ii) Risk analysis

Once the risks have been identified they should be classified in terms of:

- the likelihood of occurring;
- the impact on the project objectives should they occur.

Each risk can then be given a ranking, possibly:

- extremely unlikely;
- unlikely;
- moderate;
- likely;
- almost certain.

Note that 'will not occur' and 'certain to occur' are not classifications or we would then be considering factual certainties rather than possible risks.

Impact can also be given a ranking, possibly:

- insignificant;
- moderate;
- major;
- severe;
- catastrophic.

(iii) Risk planning

This is the process of developing responses to address, manage or control risks. The responses will consider:

- *avoiding the risk* – the Employer abandoning the project, the Contractor deciding not to tender;
- *reducing the probability* – doing something a different way or educating the people who may be subject to the risk, so the risk is less likely to occur;
- *reducing the impact* – including resources, funds and/or float in programmes to cushion the effect of the risk;
- *transferring the risk* – passing the risk to another (willing) party or an insurance company who is best able to manage or bear the risk;
- *accepting the risk* – with the full knowledge of what the risks and potential outcomes are;
- *ignoring the risk* – with the full knowledge of what the risks and potential outcomes are.

(iv) Risk tracking

Risks should be tracked throughout the project to monitor the status of each risk as the project progresses. This is normally done through a Risk Register.

(v) Risk controlling

This activity utilises the status and tracking information about a risk or risk mitigation effort. A risk may be closed or watched; a contingency plan may be involved.

(vi) Risk communicating

This is an action to communicate and document the risk at all times. Again, this will normally be done using a Risk Register.
   When considering risk it is vital to consider:

- which party can best foresee the risk;
- which party can best bear the risk;
- which party can best control the risk;
- which party most benefits or suffers if the risk materialises.

The use of a Risk Register under the ECSC

NEC3 introduced clauses proving for the use of a Risk Register within the ECC. However, it was not introduced to the ECSC, primarily as the ECSC is intended 'for use with contracts which do not require sophisticated management techniques, comprise straightforward work and impose only low risks on both the Employer and the Contractor', as the contract itself states.

Many practitioners are using Risk Registers with the ECSC. It is important to understand the purpose of a Risk Register, which is to list all the identified risks and the results of their analysis and evaluation. It can then be used to track, review, monitor and communicate risks as they arise to enable the successful Completion of the project. The Risk Register does not allocate risk – that is done by the contract; rather, it registers the risk.

The Risk Register should include a description of the risk and a description of the actions which are to be taken to avoid or reduce the risk (Figure 1.2).

There is no stated list of components of a Risk Register, but column headings should typically be titled as follows:

- *Description of risk*: A clear description of the nature of the risk, if necessary referring to other documents, such as site investigations.
- *Implications*: What would happen if the risk were to occur?
- *Likelihood of occurrence*: This provides an assessment of how likely the risk is to occur. The example in Figure 1.2 shows a forecast on a 1 (least likely) to 5 (most likely) basis, though it may be assessed as percentages, colour coding or simply 'Low' (less than 30 per cent likelihood), 'Medium' (31–70 per cent likelihood), 'High' (more than 70 per cent likelihood).
- *Potential impact*: This assesses the impact that the occurrence of this risk would have on the project in terms of time and/or cost. The example shows the assessment on a 1 (low) to 5 (high) basis.
- *Risk score*: Risk = Likelihood of something occurring × Impact should it occur.

By evaluating risk on that basis, each can be evaluated and categorised into an order of importance. This formula can only be applied to economic loss, and different standards would need to be adopted when considering risk of death or serious bodily injury.

- *Risk owner*: This identifies which party or individual owns the risk. This is allocated through the contract, the principle being that if the Contractor owns it, he is deemed to have allowed for its Price and/or time effect within his tender. If the Employer owns it, it will be a compensation event.
- *Mitigation strategy*: This registers what actions are proposed to be taken to prevent, reduce or transfer the risk. Also, who is responsible for this action, and when should the action take place?
- *Allowance in the total of the Prices*: How much has the Contractor allowed in his tender for a risk he owns?
- *Programme allowance*: How much time has the Contractor allowed in his programme for a risk he owns?
- *Employer cost allowance*: If it is an Employer's risk, how much has he allocated to cover a potential compensation event?

RISK REGISTER

Contract:.............

Contract No:.............

Contractor.............

Project Manager.............

Description of Risk	Implications	Likelihood of Occurrence (1–5) (Least–Most)	Potential Impact (1–5) (Low–High)	Risk Score	Risk Owner	Mitigation Strategy By whom? By when?	Allowance In the total of the Prices	Programme Allowance	Employer cost allowance	Risk Status	Last Updated
Discovery of unforeseen existing silo bases	Delay and additional costs in removal	2	4	8	Initially Contractor unless additional silo bases discovered then Employer as comp. event	Site Information identifies likely locations Contractor to take due regard	No allowance for unforeseen	No allowance for unforeseen	£20,000	Reducing Excavation 50% complete No additional silo bases discovered to date	15/06/2011

Figure 1.2 An extract from a typical register, including the basic requirement for a description of the risk and a description of the actions which are to be taken to avoid or reduce the risk.

- *Risk status*: This identifies whether the risk is current, and also whether it is increasing, decreasing or has not changed since it was last reviewed.
- *Last updated*: When was an additional risk identified? When was the Risk Register last updated?

In effect, the Contractor could seek to qualify his tender by including additional matters which he sees, or would like to transfer, as Employer's risks.

Risk management is essential to the success of any project and normally follows five steps:

- Step 1: Identify the risks;
- Step 2: Decide who could be harmed and how;
- Step 3: Evaluate the risks and decide on precautions;
- Step 4: Record the findings and implement them;
- Step 5: Review the assessment and update if necessary.

In order to compile the Risk Register the risks must first be listed. Then they are quantified in terms of their likelihood of occurrence and their potential impact upon the project.

The use of risk reduction meetings under the ECSC

The ECC has always had 'early warning meetings' (renamed under NEC3 as 'risk reduction meetings') where either the Project Manager or the Contractor may instruct the other to attend a risk meeting with a view to considering proposals, seeking solutions and deciding actions to reduce the notified risk.

The key word in Clause 16.2 is 'instruct'. If the Contractor calls a risk reduction meeting, he is instructing, not merely requesting, the Project Manager to attend. This clause clearly intended to promote ownership of the project by the Contractor and the Project Manager, and any issues that could affect it, together with their resolution. A risk reduction meeting need only have the Project Manager and the Contractor present, though each may instruct other people to attend if the other agrees, so in reality a number of people normally attend the meeting. The 'rule of meetings' will often apply in that the productivity of the meeting is often inversely proportional to the number of people who attend it!

The ECSC has never included provision for formal early warning/risk reduction meetings, for the reason stated in the previous section, in that it is normally a straightforward low-risk project, merely stating less formally under Clause 16.2 that:

> the Contractor and the Employer cooperate in making and considering proposals for how the effect of each matter which has been notified as an early warning can be avoided or reduced and deciding and recording actions to be taken.

However, many commercial clients adopt and include in their contracts provisions for a formal risk reduction meeting in line with the requirements of the ECC, describing the purpose of a risk reduction meeting:

- to make and consider proposals for how the effect of the registered risks can be avoided or reduced ('registered' refers to the Risk Register);
- to seek solutions that will bring advantage to all those who will be affected;
- to decide on the actions which will be taken and who, in accordance with this contract, will take them;
- to decide which risks have now been avoided or have passed and can be removed from the Risk Register.

Clearly, the purpose of a formal risk reduction meeting is actively to consider ways to avoid or reduce the effect of the matter which has been notified. In some cases, the matter can be fully resolved, but in others, as the matter may not yet have occurred, it may simply have to be 'parked' and recorded as such in the Risk Register. It is the Employer's responsibility to revise the Risk Register to record the outcome of the meeting, and to issue the revised Risk Register to the Contractor.

Note

1 Architect Denys Hinton, speaking at the RIBA, London, in 1976.

2 Design

2.1 Introduction

Design, and, more specifically, Contractor's design, is covered within the Engineering and Construction Short Contract (ECSC) by Clauses 14.3 and 20.2.

2.2 Some principles of design and design management

Good design and the management of the design process are often underrated, but critical, components in the success or failure of construction projects.

It is imperative that the Employer initially understands what he wants and is able to communicate that clearly to the Designer and other parties in the form of a clearly structured brief. The Designer can then convert that brief into a practical and economic solution which provides the deliverables of meeting quality standards, within budget and available time constraints.

Whilst quality and financial constraints tend to be uppermost in Employers' minds, it is critical that sufficient time is allowed for the design to be developed and alternatives considered, then the often time-consuming process of submission and awaiting regulatory approval, before making final decisions.

In the European Union, consideration must also be given when designing projects in the public domain to time scales to comply with *Official Journal of the European Union (OJEU)* regulations.

Key points in effective design management are as follows:

- Employers should always be clear as to their objectives, priorities and operational requirements, including the needs of all stakeholders, before appointing Designers, and ensure that these are clearly stated in their brief. The initial stages are the most important part of any project, and a well-informed Employer is the critical foundation to the successful delivery of the project within the budget, agreed time scale and to the required standards.
- Allocate sufficient time and resources to establish the Client's design quality aspirations, ensuring that design time is not reduced in order to recover time lost in the programme for other reasons. Good design always takes time. Time allocation should also include the need for external regulatory approvals over which the Employer and his Designers may have little or no control.

- Appoint someone responsible for taking the lead on design issues and coordinating the work of the various parties, again including regulatory bodies.
- Continue to communicate design requirements throughout the design, procurement and construction stages.
- Good design should always consider the most economical solution, both in terms of initial cost, but also life cycle cost, including future maintenance and removal and replacement; therefore value engineering and continuous design reviews are integral to the process.
- Good design should consider not only aesthetic and structural features, but also functionality and efficiency, build quality, accessibility, sustainability, whole life costing and flexibility in use.
- Employers should always be clear as to what funding they have available for the design process and for what is to be designed. That funding should always be realistic.
- A candid and honest, but formal post-Completion audit should always be carried out to ascertain the successes and failures of the project from which lessons can be learned and carried through to the next project.

2.3 Selection of the design team

There are several methods of selecting the appropriate design team for a particular project, including, particularly for a high-profile project, design competitions, which primarily select specific design ideas or outline designs for a project, rather than the individual design team members.

The key issue with design competitions is that the Employer must have a clear vision of what he requires in order that design teams will have clear criteria from which to create and submit their designs.

Design competitions may be appropriate where there is either a unique problem that will benefit from a wide range of design approaches being explored (along with likely considerable public interest or resistance – which may be the case on a major new public building) or where the competition promoter wishes to encourage the development of new talent.

Alternatively, one can focus on the individual Designers and their skills and reputation in designing similar projects. Again, it is vital that not only aesthetic qualities are considered, but also the design team's ability to perform within the time constraints and budget, i.e. to manage the design.

With the European Union, all public sector appointments, irrespective of the Client's preferred nature of competition or reference to any other guidance on design competitions, must be consistent with EC procurement rules in terms of process and outcome.

Whichever process is used for selecting the design team, there must be clear and objective criteria on which to select and appoint the successful Designers and this should consider:

- the balance between quality, cost and time-related issues;
- the technical ability of the design team to deliver the project on time and within budget, as well as to the required standards of quality;
- the financial standing of the Designers, including insurance requirements. The financial failure of a Designer can be catastrophic to the successful Completion of the project.

2.4 Contractor's design

The NEC3 is unique amongst construction contract families in that, whilst most other contracts provide for portions or all of the design to be carried out by the Contractor, they all use a separate contract for design and build where the Contractor has full responsibility for design, whereas the Engineering and Construction Contract (ECC) and ECSC use the same contract whether or not design is to be carried out by the Contractor, as any design criteria and obligations are defined within the Works Information.

In reality, there is no need for a separate design and build contract as within the ECC and ECSC the clauses which cover early warnings, programme, payment and change management are the same whether or not the Contractor has design responsibility.

The 'default position' within the ECSC is that all design is carried out by the Employer, but if any or all of the design is to be carried out by the Contractor this is assigned to him within the Works Information. Clause 20.2 refers to the Contractor needing to obtain acceptance from the Employer of anything that he has designed (Figure 2.1).

The most important advantage with design and build is that the Contractor is wholly responsible for the design. This 'single-point' responsibility is very attractive to Employers, particularly those who do not have the knowledge or experience to appoint design consultants, or who may not be interested or able to distinguish the difference between a design fault and a workmanship fault. However, it is vital that the Employer is able to detail his requirements clearly.

This single-point responsibility also means that the Contractor is not relying on other firms (e.g. Architects) for the execution of design or for the supply of information.

The design and build process increases the opportunities for harnessing the benefit of the Contractor's experience during the design stages of the project.

Many of the developments in procurement processes, and much of the work in the field of study known as 'buildability', have had this purpose in mind. The benefits of the integration of Designers and builders are more economically constructed projects, as well as a more economic and effective production process.

The technical complexity of the Client's process may lead to solutions which are more akin to engineering than to building. While this does not mean that design and build is unsuitable, it is likely that in such circumstances the use of other forms of procurement will be favoured.

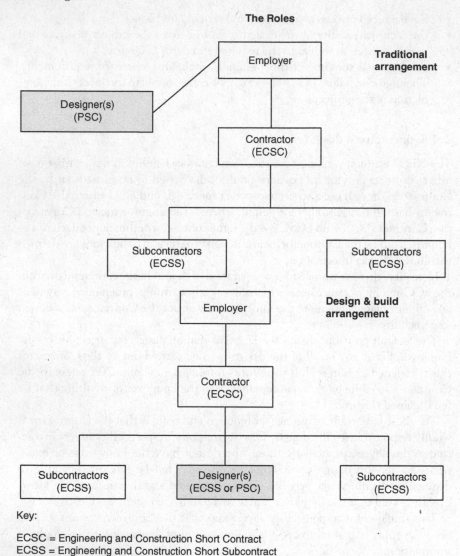

Figure 2.1 The traditional arrangement versus the design and build arrangement.

The reasons for allocating some or most of the design to the Contractor can then be summarised as follows.

Advantages

* Single-point responsibility, so the traditional design/construction interface and the risks associated with it are transferred to the Contractor. Whether a

Defect is a design fault or a workmanship fault, it is the responsibility of the Contractor to correct that Defect.

- It is not necessary for the design and build Client to be an 'expert' Client and appoint and manage a group of consultants. He engages the Contractor as a single point of responsibility to provide that expertise.
- The management of the design risk by the Contractor can result in greater certainty of the time, cost and project objectives being met.
- The procurement period is shorter as there is no requirement for the Employer to complete the design and prepare bills of quantities prior to appointing the Contractor.
- The project can be completed on site quicker as design and construction can overlap.
- The Contractor can utilise his experience and expertise in using preferred methods of construction to minimise time and cost.
- As the Contractor has responsibility for design and construction, the design should be more rational and buildable.

Disadvantages

- The Contractor designs and builds to a price commitment, therefore where there is a conflict between aesthetic quality and ease of fabrication, the requirements for fabrication will override.
- The Employer has reduced control of design and quality as these are passed to the Contractor at an early date.
- The Employer has reduced control of change costs as there is no detailed bill of quantities to relate to.
- It may not be good for complex projects because of the more comprehensive pre-contractual documentation which may be required by the Client.
- Whilst the procurement period may be shortened, the Contractor must be given more time to prepare his tender.
- It is expensive for the Contractor to prepare his tender as he needs to consult with Designers as well as to price the tender.

If the Contractor is to design all or part of the works, the Employer must create a clear and unambiguous set of requirements, clearly defining what he requires the tendering Contractors to price and eventually to carry out. The Contractor responds with his proposals, which will include production as well as design work.

The Contractor's design input varies from one contract to another, ranging from the mere development of a detailed design which was provided by the Employer to a full design process, including proposals, sketch schemes and production information.

There will often be some negotiation between the Employer and the Contractor, with the aim being to settle on an agreed set of Contractor's proposals. These proposals will include the contract price, as well as the manner in which it has been calculated.

Once the Employer's requirements and the Contractor's proposals match, the contract can be executed and the Contractor can implement the work.

One reported disadvantage of design and build is where there is a conflict between aesthetic quality and ease of construction. As the Contractor is also designing the project, the requirements for construction will override.

A further criticism has been that a design and build Contractor will put in the minimum effort with regard to the design in an effort to secure the contract.

These two criticisms suggest strongly that quality, particularly design quality, will suffer under this procurement process. However, this is not a valid criticism of design and build as a procurement form, but rather of the people who advise Clients and prepare the Employer's requirements and tender documents.

When selecting a Contractor, the Employer may approach design and build Contractors who will prepare the design and submit their tenders based on a 'shopping list of requirements', or in many cases will employ his own Designers to carry out initial designs of major elements of the Works, which would then form the basis of the Employer's requirements, with the Designers transferred to the Contractor by a novation agreement.

2.5 Incorporating Contractor's design into the ECSC

Establishing the Employer's Works Information

In the ECSC, the Contractor designs the parts of the works which the Works Information states he is to design.

It is clear, then, that the Works Information should clearly show what is to be designed by the Contractor and this could consist of detailed specifications or some form of performance criteria and certain warranties.

When preparing the Works Information, which includes Contractor's design, it is important that the right balance of information should be considered. If it is too prescriptive it will lead to all the tendering Contractors submitting similar designs and prices. If the Works Information is not detailed enough, then none of the tendering Contractors will produce a design which is as the Employer intended.

As an example, the Employer wishes to invite tenders under the ECSC for the design and build of 8 No. houses. The Employer has certain essential requirements, but wishes to attract some innovation in that Contractors will interpret the shape, size and appearance of the houses in different ways subject to the approval of external bodies such as the local authority planners and other legislative bodies.

The Works Information should therefore comprise the following groups:

- *Essentials* – requirements that the Employer must have. For example, if the Contractor is to design and build houses then the number of houses and the mix of types of houses, for example number of bedrooms, should be clearly stated. Any other essentials should also be clearly stated; otherwise a tenderer may not include them.

- *Preferences* – requirements that the Employer would prefer to have. For example, the Employer may have a preference for a certain heating system, but the Contractor may offer an alternative which is more efficient and cheaper to run.

  The Employer may also state within his preferences that 'equal and approved' may be acceptable subject to the Contractor proving that it truly is 'equal' and also that he is prepared to approve it. As part of this proof the Contractor should consider issues such as life cycle costing. For example, if the Contractor can prove that the alternative heating system is equally efficient and easy to operate, but in the longer term the alternative requires a higher level of maintenance, then the Employer will need to be made aware of this requirement. It is suggested that the preferences should not apply to critical aspects of the Project.
- *Innovations* – something the Employer may not have thought of that the Contractor believes could win the tender for him!

  This is open to the Contractor to provide as added value to his particular bid. For instance, one Contractor may offer a children's playground, or particularly 'green' design/construction over and above the requirements of the Works Information and/or the relevant legislation. If a Contractor feels he can exceed the Works Information he should submit his proposals which comply with the document (a compliant bid) but list separately how he could exceed it and any time and/or Price implications.

  With regard to the Contractor's proposals, he will wish to demonstrate that his bid is compliant with the Works Information, though it is important that he is not prescriptive about exactly what he will provide, as his design may evolve and develop as the construction progresses. He may find new Equipment and Materials, or he may wish to use some monies to offset any problems that may arise.

If one is constructing the Works Information for the design and build of the 8 No. houses, then one should consider the following.

Quality

What are the Employer's overall requirements?

- The size and use of the project. If no specific size of buildings is known, details of exactly what the Employer intends to use the proposed building for and expected number of occupants, e.g. number of bedrooms;
- Number of floors;
- Any specific room layouts;
- Any specific requirements, for example, kitchen and bathroom layouts or specialist suppliers or subcontractors to be considered/appointed;
- Need for future flexibility and expansion;

- Specific heating requirements, lighting levels for various parts of the building, heating/air conditioning requirements. Running costs and energy conservation requirements;
- Car parking and external landscaping requirements;
- Maintenance, running costs and life cycle cost factors. How long are the houses expected to last?

Time

When will the Employer need the project completed?

- Are the Starting Date and/or Completion Date critical to the Employer?
- Consider pre-tender activities such as planning consents, preparation of tender documents and also other operations which need to be carried out before the Contractor can be appointed and start on site.

Cost

- How much will be paid?
- What is the budget for the project? How restricted is that budget? Is there a cost plan and expected cash flow projections for the project? How long will it take to prepare one?
- Is the project being wholly funded by the Employer or is partial funding coming from another source, for example grant aid? How much influence do the funders have in terms of the design and construction of the project?

From these initial considerations the documentation to be prepared for tender can be established:

- What drawings are to be included within the Works Information? Broad sketch plans, showing size and general layout only? Or detailed design drawings?
- Site Information. How detailed will it be? Who is responsible for accuracy?
- Who will apply for outline and later detailed planning consents? Possible delays with consents? Revised applications due to changes in design suggested by Contractor?
- Performance specifications? Or specification of Works and Materials? Outline specifications or detailed at this stage?
- Any major products, specialist or supply items over which the Employer wishes to exercise some choice or control. Nomination is difficult with design and build and creates complications.
- Employer to provide an outline time programme, or left to the Contractor to submit with his tender?
- Sectional Completions? Split responsibilities and Insurances.
- Constraints: access, space, working hours, sequence of working.

- Which Main and Secondary Options are to be used?
- Design and construction liability. Any special insurance requirements or performance bonds? Design warranties?

From the above it can be readily seen that it is vital that the Employer defines how much information and detail is required in the returned tender and it is crystallised within the Works Information. For example, a full set of drawings and comprehensive specification would not normally be expected from tenderers at tender stage and the scope and limitation of these documents must be defined.

The Contractor's design

In simple terms, the Contractor's design should be fully compliant with the Employer's Works Information, and the Contractor cannot start work that he has designed until the Employer has accepted that the design complies with the Works Information (Clause 20.2). If the Employer's Works Information is detailed, the Contractor's design details would probably be detailed; if it is vague, the Contractor's design details may also be vague.

The Contractor's design details within his tender should incorporate the following:

- plans, elevations, sections or typical details;
- information about the structural design;
- outline drawings of incorporated services;
- specifications for materials and workmanship;
- any alternative solutions departing from the brief;
- a method statement and construction programme;
- the price;
- a staged payment schedule (if appropriate);
- proposed main Subcontractors and Suppliers.

If the Contractor's design proposals subsequently fall short of the Employer's Works Information then the Employer's Works Information will prevail.

Employer's Works Information versus Contractor's design

The ECSC is unlike most other construction contracts in that it does not provide a priority or hierarchy of documents, but one can 'read into' the various clauses to determine which takes precedence in the event of an inconsistency.

A question to be considered is, if the Contractor has full design responsibility and there is an inconsistency between the Works Information provided by the Employer and the design provided by the Contractor, which takes precedence?

The answer can be found in Clauses 11.2(3) and 20.2:

- A Defect is defined in Clause 11.2(3) as a part of the Works which is not in accordance with the Works Information.
- Under Clause 20.2, the Contractor does not start work which he has designed until the Employer has accepted that the design complies with the Works Information.

Clearly the Employer's Works Information takes precedence.

2.6 Acceptance of design

Under Clause 20.2, the Contractor does not start work which he has designed until the Employer has accepted that it complies with the Works Information.

The Contractor is therefore required to submit his design as the Works Information requires to the Employer for acceptance. The particulars of the design in terms of drawings, specification and other details must clearly be sufficient for the Employer to make the decision as to whether they comply with the Works Information and also, if relevant, the applicable law.

The Works Information may stipulate whether the design may be submitted in parts, and also how long the Employer requires to accept the design. Note that in the absence of a stated time scale for acceptance of design, the 'period for reply' will apply.

Many Employers are concerned that they have to 'approve' the Contractor's design and therefore they are concerned as to whether they would be qualified, experienced and also insured to be able to do so. Note the use of the word 'accept' as distinct from 'approve'. Acceptance denotes compliance with the Works Information; it does not denote that the design will work, that it will be approved by regulating authorities, or that it will fulfil all the obligations which the contract and the law impose, therefore performance requirements such as structural strength and insulation qualities do not need to be considered.

A reason for the Employer not accepting the Contractor's design is that it does not comply with the Works Information. The Contractor cannot proceed with the relevant work until the Employer has accepted the design.

Note that, under Clause 14.3, the Employer's acceptance of the Contractor's design does not change the Contractor's responsibility to Provide the Works or his liability for his design. The Employer is purely accepting the design in that it complies with the Works Information.

The Contractor will also probably need to obtain approval of his design from others where necessary. This will include planning authorities and other third-party regulatory bodies.

This is an important aspect as, if the Employer accepts the Contractor's design but following acceptance there are problems with the proposed design, for example in meeting the appropriate legislation or the requirements of external regulatory bodies, this is the Contractor's liability.

2.7 Intellectual Property Rights

The question of Intellectual Property Rights (IPR), which includes copyright, patents and trademarks, needs to be considered carefully where the Contractor is designing all or part of the Works.

The design and construction of a building or other structure are unique and, unless the contract states otherwise, even though the owner pays for the design work, he does not automatically own the copyright to the design and the design documents. The Designer retains ownership and control over the design and the various documents, based on copyright law and the terms of agreement between the Client and the Designer.

This copyright issue can be very complex as it only applies to that which was originally conceived by the Designer and therefore much of what is included in the design may not be subject to copyright, for example the layout of kitchen units or sanitary appliances within a bathroom.

Copyright also extends to copying the design, photocopying and distributing drawings and other documents.

In the event that copyright is breached, the owner of the copyright may obtain an injunction preventing further breaches and, if necessary, may seek impounding of Materials, and also damages, loss of profits and legal fees. If a building design copyright has been breached the Client may be prevented from commencing or continuing with the works.

2.8 Collateral warranties

A collateral warranty is an agreement associated with another contract – collateral meaning 'additional but subordinate' or 'running side by side'. The collateral warranty is entered into by a party to the primary contract, for example a Contractor or a Subcontractor and a third party who is not a party to the primary contract but who has an interest in the construction project.

Collateral warranties are a relatively recent feature of contracts, coming into use in the past 20 years following some notable court cases in which it was judged that it was not possible to recover damages for negligence in relation to Defects in construction projects as it was held to be economic loss, which is not recoverable in tort.

Consequently, such claims had to be brought as claims for breach of contract, where economic loss is recoverable. It therefore became necessary to establish a contract between parties involved in a construction project, such as Designers or Subcontractors, and parties such as funders and present and future purchasers, who are not parties to the primary contracts but who have an interest in the project, hence the 'collateral warranty'.

The NEC3 contracts do not provide for collateral warranties, though it is worth some commentary on their use and how they could be included specifically within the ECC.

Collateral warranties are different from traditional contracts, in that the two parties to the warranty do not have any direct commercial relationship with each other. They are useful for creating ties between third parties, but the main disadvantage of collateral warranties is the expense of providing them, particularly where there are many interested third parties such as with housing and commercial properties. When one also considers that parties can change their names or business status during the life of a warranty then even a medium-sized project can involve 30 or more warranties.

Different parties will also have different requirements for collateral warranties. Therefore the precise terms of the collateral warranty agreement need to be considered in each case carefully so that they are drafted to provide the required protection.

Collateral warranties are executed as a 'deed', also referred to as 'under seal', which differs from a simple contract in two respects:

- First, in English law the time after a breach of contract has occurred within which one party can sue another is twelve years, whereas the time for a simple contract executed 'under hand' is six years.
- Second, whereas under English law consideration is needed for a contract to be effective, this is not the case with a deed. For a company to execute a deed effectively, the document must be signed by two directors or a director and the company secretary, unless the Articles of Association contain some other requirement.

The basic ingredients of a collateral warranty are usually as follows.

(i) Definitions

There will be a brief description of the Works and the collateral warranty. The 'contract' is the contract to which the collateral warranty agreement is collateral and the date should be specified. Since the obligations of the Contractor to the Beneficiary parallel the obligations of the Contractor to the Developer under the contract for the works, the Beneficiary should always ask to see a copy of the contract. If, for example, there are restrictions on the Contractor's liability in the principal contract, that might affect the rights of the Beneficiary under the collateral warranty.

A copy of the principal contract may be attached to the collateral warranty agreement.

(ii) The warranty

The Contractor gives warranties to the Beneficiary that he has exercised and will continue to exercise reasonable skill and care in his obligations to the Employer under the principal contract, including compliance with any design obligations, performance requirements and correction of Defects.

The warranty should contain a statement that the Warrantor shall have no greater liability to the Beneficiary than he would have under the primary contract.

The Contractor also does not have any liability to the Beneficiary if he fails to complete the works under the contract on time, as that will usually be covered by delay damages under the principal contract.

There should be a statement that the Warrantor shall have no liability under warranty in any proceedings commenced more than 12 years after the date of Completion of the Works.

(iii) Naming of the parties

The named parties can be, for example, the Contractor and the end user, a Sub-contractor and the Employer, or a Consultant and the Employer.

The party that gives the warranty is known as the Warrantor; the other party is known as the Beneficiary.

(iv) Prohibited Materials

The Contractor may be required to give a warranty that he will not specify or use materials which are known to be prohibited or deleterious to health and safety.

In addition, the use of deleterious Materials may contravene relevant legislation or regulations as well as being expressly prohibited in the contract between the Developer and the Contractor.

(v) Copyright

Normally, the Beneficiary under a collateral warranty agreement will have a right to use designs and other documents prepared by the Contractor but only in connection with the project for which those design documents are prepared. Copyright to these documents remains with the Contractor.

(vi) Insurances

There is normally a requirement for the Warrantor to maintain insurances, including Professional Indemnity (PI) insurance where applicable for a stated minimum level of cover and for a period of 12 years after Completion of the Works.

(vii) Assignment

There is normally a provision allowing the Beneficiary up to two assignments of the warranty.

(viii) Step-in rights

Step-in rights can allow a funder to take over the Employer's role and become the Employer under the primary contract. This is particularly important where the Employer is unable to pay the Contractor, the Contractor having to notify the Beneficiary before he terminates the contract, thus allowing the Beneficiary to 'step in' and take the Employer's place, with an obligation to pay the Contractor any money that is outstanding. The Beneficiary does not have any obligation to step in, only the right to do so. As the effect of this clause is to vary the principal contract it is important that, when the collateral warranty contains such a clause, the Employer should be a party to that document.

(ix) *Law and jurisdiction*

It is important to specify the law of the contract and the warranty and the jurisdiction of the courts that will have jurisdiction.

2.9 Novation of Designers

Whilst the principle of design and build agreements is that the Employer prepares his requirements and sends them to the tendering Contractors and they prepare their proposals to match the Employer's requirements, in reality in nearly half of design and build contracts the Employer has already appointed a design team which prepares feasibility proposals and initial design proposals before tenders are invited. Outline planning permission and sometimes detailed permission may have also been obtained for the scheme before the Contractor is appointed.

Each Contractor then tenders on the basis that the Employer's design team will be novated or transferred to the successful tendering Contractor who will then be responsible for appointing the team and completing the design under a new agreement.

This process is often referred to as 'novation', which means 'replace' or 'substitute', and is a mechanism where one party transfers all its obligations and benefits under a contract to a third party. The third party effectively replaces the original party as a party to that contract, so the Contractor is in the same position as if he had been the Employer from the commencement of the original contract. Many prefer to use the term 'consultant switch', where the design consultant 'switches' to work for the Contractor under different terms, as a more accurate definition.

This approach allows the Employer and his advisers time to develop their thoughts and requirements and consider planning consent issues. Then when the design is fairly well advanced, the Designers can be passed to the successful design and build Contractor.

The NEC3 contracts do not include pro forma novation agreements, but, if used, it is critical that the wording of these agreements be carefully considered as there are many badly drafted agreements in existence. Novation can be by a signed agreement or by deed. As with most contracts, there should be consideration, which is usually assumed to be the discharge of the original contract and the original parties' contractual obligations to each other. If the consideration is unclear, or where there is none, the novation agreement should be executed as a deed.

In many cases the agreement states briefly (and very badly) that from the date of the execution of the novation agreement, the Contractor will take the place of the Employer as if he had employed the Designers from the beginning. The document states that in place of the word 'Employer' one should read 'Contractor'; however, the issues of design liability, inspection, guarantees and warranties may not always apply on a back-to-back basis, so one must take care to draft the agreement in sufficient detail and refer to the correct parties.

In practice, the best way is to have an agreement drafted between the Employer and the Designer covering the pre-novation period, and a totally separate agreement drafted between the Contractor and the Designer for the post-novation period.

When a contract is novated, the other (original) contracting party must be left in the same position as he was in prior to the novation being made. Essentially, a novation requires the agreement of all three parties.

As this book is intended as a guide for NEC3 practitioners worldwide, it is not intended to examine specific legal cases within the UK, as they may not apply on an international basis. There is certainly case law available in terms of design errors in the pre-novation phase being carried through as the liability of the Contractor in the post-novation stage, and also the accuracy and validity of Site investigations and the Employer's and the Contractor's obligations in terms of checking and taking ownership of such information.

2.10 Fitness for purpose versus reasonable skill and care

In order to consider the design responsibilities of the parties, we should first consider the principles of 'fitness for purpose' and 'reasonable skill and care'.

If one considers a building project as an example, 'fitness for purpose' means producing that building, part of a building or a product fit for its intended purpose.

This is an absolute duty independent of negligence. In the absence of any express terms to the contrary, a Contractor or Subcontractor who has design responsibility will be required to design 'fit for purpose'.

'Reasonable skill and care' means performing to the level of an ordinary competent person exercising a particular skill. The criterion is that it is not possible or reasonable to judge someone's professional ability purely by results; for example, a lawyer whose client is found guilty is not necessarily a poor lawyer, and a surgeon whose patient dies is not necessarily a poor surgeon. The question to ask is 'did that professional apply the correct level of skill and care in carrying out his duties?'

If one then applies the same criterion to a designer, this is normally achieved by the Designer following accepted practice and complying with statutory requirements and codes of practice, but if a new construction technique is involved, the duty can be discharged by taking advice from consultants and warning the Employer of any risks involved. A further qualification on the Designer's liability is the 'state-of-the-art' defence, meaning that a Designer is only expected to design in conformity with the accepted standards of the time.

In the absence of any express terms to the contrary, a Designer will normally be required to design using 'reasonable skill and care'.

Design and build contracts can impose a higher standard than reasonable skill and care, i.e. fitness for purpose. This obligation resembles a seller's duty to supply goods which are fit for their intended purpose.

The 'default' position within the ECSC is that any design carried out by the Contractor must in all respects be fit for purpose. As stated above, this means producing a building, part of a building or a product fit for its intended purpose. This is an absolute duty independent of negligence.

In the absence of any express terms to the contrary, a construction contract which includes design requires the Contractor to design 'fit for purpose'.

The Joint Contracts Tribunal (JCT) 2011 Design and Build Contract states that:

> the Contractor shall in respect of any inadequacy in such design have the like liability to the Employer, whether under statute or otherwise, as would an Architect or, as the case may be, other appropriate professional Designer holding himself out as competent to take on work for such design who, acting independently under a separate contract with the Employer, has supplied such design for or in connection with Works to be carried out and completed by a building Contractor who is not the Supplier of the design . . .

This is the JCT version of 'reasonable skill and care'!

'Reasonable skill and care' means designing to the level of an ordinary competent person exercising a particular skill. This is normally achieved by the Designer following accepted practice and complying with standards and codes of practice.

Note the 'state-of-the-art' defence that could be applied in the case of a new type of material or a new and as yet untried construction technique, in which case the duty can be discharged by taking advice from consultants and warning the Employer of any risks involved.

In the absence of any express terms to the contrary, a contract for design requires the Designer to design using 'reasonable skill and care'.

The Contractor is not liable for Defects in the works due to his design so far as he proves that he has used reasonable skill and care to ensure that his design complied with the Works Information and any other required criteria.

From the above, it is hopefully clear that Employers should take advice, as 'fitness for purpose' is a very onerous requirement for a Contractor with design responsibility for a typical project covered by the ECSC, so it may be prudent to consider relaxing the Contractor's design obligation to that of 'reasonable skill and care'.

2.11 Dealing with planning permission

The responsibility for obtaining planning permission can lie with the Employer or can be assigned to the Contractor.

In some cases, the Contractor may be responsible for gaining outline planning for the project. However this should be a two-stage process so that if outline planning permission is not obtained the second stage would not proceed.

It is more usual for the Employer to gain outline planning permission and the Contractor to be responsible for gaining full and detailed planning consent.

3 Time

3.1 Introduction

Time is covered in the Engineering and Construction Short Contract (ECSC) by Clauses 30.1 to 31.1.

Unlike the Engineering and Construction Contract (ECC), the ECSC does not have core clauses expressly requiring the Contractor to submit a programme for acceptance, though the Works Information (Section 5) allows the Employer to state whether a programme is required, what form it is to be in, what information is to be shown on it, when it is to be submitted and when it is to be updated. Also the Employer can state what the use of the works is intended to be at their Completion, as defined in Clause 11.2(1).

Most commercial users of the ECSC tend to 'copy and paste' relevant sections from ECC Clause 31.2 into the ECSC so that the Contractor is required to submit a programme in a prescribed format, though there is still no express requirement for the programme to be accepted.

3.2 General principles of planning and programme management

A successful construction project is considered by Employers and their Consultants to be one which has been completed on time, to the required quality and also within budget; and by Contractors and Subcontractors as one which has been completed on time, to the required quality and making them a profit! Clearly, apart from their different perspectives on financial outcome, the objectives of all contracting parties are well matched.

When considering Completion on time, various surveys carried out in recent years have shown that only about 33 per cent of contracts are completed within the contract period, one-fifth overrun by 40 per cent and a significant number overrun by more than 80 per cent!

It is very important, even on a straightforward low-risk contract for which the ECSC was designed, that the parties are able to set out their intentions in terms of time scales and the resources they intend to use, so that progress can be monitored and meaningful assessments made in terms of delays, including, but not limited to, those attributable to compensation events.

There are many factors which can cause delays even on what, in theory, is a straightforward project:

- delays/errors in design, often exacerbated by poor communications between design and construction teams;
- inadequate site investigations. Clients often believe that Contractors should be responsible for assessing ground conditions based on poor and inadequate information provided by them;
- inability to give timely access or possession;
- non-availability of resources, by all parties;
- poor performance of parties;
- Client changes, compounded by the number or timing of these instructions;
- weather and other seasonal problems;
- insolvency of a party;
- unrealistic construction period set by Employer.

Clearly, matters need to be improved by the parties and between the parties, but many of these factors can be avoided or at least reduced by the parties engaging in proper and effective planning and programming of the Works aligned to effective communication between the parties.

3.3 Planning and programming

The terms 'planning' and 'programming' are used interchangeably; however, their meanings are a little different.

What is planning?

Planning may be defined as 'the deliberate consideration of all the circumstances concerned with a project in order to evolve the best method of achieving a stated objective'. This involves defining the project, setting the overall duration and identifying risks.

Planning is a vital element in any construction process, without which it is impossible to envisage the successful conclusion of any project.

Planning considers:

- what to do;
- when to do it;
- how to do it;
- what resources are required to do it;
- who is going to do it;
- how long it will take to do it.

The motto is: Plan your work . . . And work to your plan!

Planning may be considered on a long- or short-term basis. Long-term planning includes the development of a master plan for the whole project, whilst short-term planning is a much more detailed operation, which considers only the next three to four weeks of operations, or a particular element of the work. It is essential that each short-term plan should overlap the other by one to two weeks to allow for past and future activities to be considered.

The planning process considers each operation in turn, analysing the order and sequence of each operation, its dependency on previous operations, the Contractor's and Employer's resources to carry it out and the availability of those resources.

It is easier to track progress of operations using short-term, rather than long-term, planning.

There are three planning stages: (1) pre-tender; (2) pre-contract; and (3) project planning.

Pre-tender planning

This covers the planning considerations during the preparation of a cost estimate and its conversion into a submitted tender Price.

The process enables the Contractor to:

- establish a realistic construction period on which the tender may be based;
- consider the most effective economical construction methods, using the most effective resources;
- assist the build-up and pricing of contract establishment charges and Equipment expenditure.

This is usually carried out in broad outline format as the Contractor may not be successful in his tender.

Typical information which may be gathered and used at the tender planning stage includes:

- a brief description of the project and its location, details of the Employer and the various consultants, key companies, third parties and individuals;
- details of the local planners and any planning issues or restrictions;
- access to site for delivery and collection of equipment, plant and materials and also removal of waste;
- details of existing services and the location of new and existing connection points, and details of the relevant service providers;
- general details of the geography and topography of the area, past use, ground investigations, groundwater levels and other local knowledge;
- location of tips and Suppliers, including quarries;
- availability of labour, equipment, plant and materials;
- details of possible Subcontractors;
- typical weather conditions.

This information is then collected together with information given to the Contractor at tender stage to develop a pre-tender plan. Time-based elements can then be transferred to a pre-tender programme, usually in a bar chart format which, together with the method statements, then aids the estimator in preparing the tender and assessing the need for establishment charges and major multi-use Equipment such as craneage, scaffold and skips.

Employers should instruct Contractors to submit at least outline programmes with their tenders, though it must be appreciated that as the Contractor may not be successful with his tender it will serve only as an indicative outline of how the Contractor would carry out the works should he be successful.

Normally a contract will provide for a Contractor to submit a first programme for acceptance within a specified number of weeks of the contract coming into existence.

Pre-contract planning

This covers the planning considerations once the contract has been awarded. It is essential that this planning stage takes place prior to commencing work on the Site.

The process enables the Contractor to:

- establish a clear strategy for carrying out the works;
- comply with contract conditions and requirements;
- establish a construction sequence on which the master programme may be based;
- identify key dates by which parts of the work are required to be completed or the Employer is to provide information or resources to enable the Contractor to complete work to the programme requirements;
- enable the assessment of contract budgets and cash flow forecasts;
- schedule key dates for procurement of key materials and Subcontractor tendering, placing orders and commencement;
- schedule pre-contract meetings.

Project planning

This covers the planning which is required to be implemented in order to maintain control and ensure that the project is completed on time and within the cost limits established at the tender stage.

The process enables the Contractor to:

- monitor and update the master programme on a regular basis;
- optimise and continually review resources;
- keep progress under review and report on variances.

What is programming?

Programming is drawing all the planning considerations together into a present-able graphical format such as a bar chart and supporting information and tables which show the various operations, their durations, relationship between the operations, critical path and float, and the required resources so that it can be used as a project reporting and control mechanism, and form the basis of progress reporting to the Employer.

Programming is an essential part of any construction contract, but whilst other contracts often pay only a passing interest to the programme, the NEC3 family sees it as a vital tool from which to:

- at tender stage, assess the Contractor's ability to carry out the project;
- measure and report the Contractor's actual progress against that planned;
- assess the effect of compensation events upon Completion and Price;
- ensure the Employer and the Contractor take appropriate action when required.

3.4 Preparation of a programme

The process of writing a programme normally consists of three stages.

(i) Identifying the various operations to be carried out

The operations may initially be written as a list, identifying each operation in chronological order. That order may depend on natural construction sequence, constraints to access and resource availability.

(ii) Considering the duration of each operation

The duration of each operation can only be a forecast rather than an exact science as various issues have to be considered, such as length of working days, differing productivity outputs, changes, resource availability, working conditions, changing weather and equipment breakdowns, which can affect forecast durations. The Contractor will then make appropriate adjustments as the work progresses.

(iii) Considering the relationship between each operation

This will involve inserting links between operations which have a dependency on each other, for example 'start to start', 'finish to start', 'finish to finish'. From this can be established what the logical path is in order to execute the various operations, operations that need to start at the same time, what operations have to be completed before another operation can start and what operations must finish at the same time.

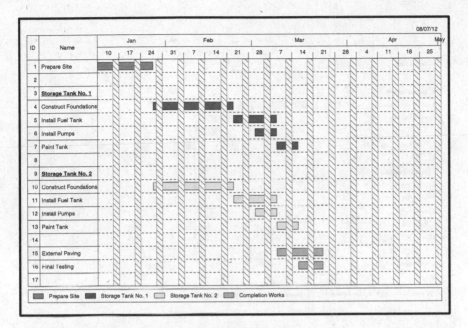

Figure 3.1 Programme with operations unlinked.

Figure 3.1 shows a simple programme for the construction of two storage tanks. One can interpret from the programme that there is an initial period for preparing the Site; the construction of the foundations to both tanks then commences once the Site has been prepared.

The construction of the tank foundations, the installation of the tanks themselves and the installation of the pumps and tank painting for both tanks then progress concurrently. Whilst the tanks are being painted, the external paving around the tanks can then progress, then the final operation, final testing, will progress, with Completion of this operation coinciding with Completion of the external paving, and thus Completion of the works.

The problem is that, without any links, the dependency of the operations upon each other, whilst implied from their positioning within the programme, is not definitive. Therefore if, for example, the construction of the foundations to Tank No. 1 is delayed, either by the Contractor or the Employer, the effect on other operations can be inferred but is not stated definitively.

Figure 3.2 includes links which show the logic and dependency of each operation. Therefore, if the construction of the foundations to Tank No. 1 is delayed, it is on the critical path and therefore the delay to the foundations will have a direct delay effect on all the following operations.

The term 'programme' is not defined in law or within contracts. The dictionary definition normally states that a programme is 'a plan or schedule of activities to be carried out'. The ECC does not prescribe the form the programme submitted for acceptance must take; it would normally be a bar chart programme.

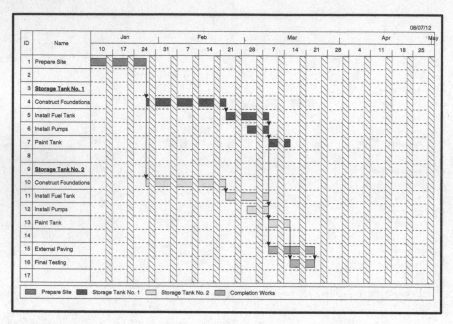

Figure 3.2 Programme with operations linked.

3.5 Types of programme

Before considering in detail the requirements of the ECC, it is worth reviewing types of programme in common use.

(i) Bar charts

Any programme, at any stage of a tender or project, is normally written as a bar chart or Gantt chart (Henry Laurence Gantt, 1861–1919) showing operations which may occur either on or off Site, including their sequence and duration, the start and finish of the bar depicting the start and finish date of the operation. Linked bar charts, as described previously, are the usual form of programme in common use on construction contracts. They are often supplemented by other information.

Advantages of bar charts

- The simple format allows them to be readily constructed; they are user-friendly and understood by most practitioners.
- They illustrate the specific activities and their durations and, during the progress of the work, actual versus planned progress can be clearly shown.
- They can be used at all stages of the planning process through pre-tender, pre-contract and contract planning. They assist in showing the relationship

between the pre-tender programme, the master programme and short-term programmes.

- They clearly relate to the construction sequence. The use of linking on bar charts aids the overlapping of related activities and can enable the critical path to be identified.
- They are easily updated at weekly and monthly time intervals for review purposes and progress reports.
- Key milestone symbols may be introduced to highlight critical dates with regard to key contract stages and information requirements.
- Resources may be readily related to the rate of working – labour histograms, value forecasts (value/time) and cumulative labour and Plant forecasts in the form of project budgets.
- The programme forms the basis of financial forecasting for both the Client and the Contractor as a cash flow forecast can be prepared by pricing the programme.

Disadvantages of bar charts

- Without the use of links they cannot properly depict the relationship and dependency between activities; for example where an activity must be wholly or partially complete before the next can start.
- Complex interrelationships cannot always be clearly shown.
- They may give too simple a picture, which may be misleading.

(ii) Elemental trend analysis (line of balance) charts

Elemental trend analysis (line of balance) charts, as with many programming methods, originated from the manufacturing industry and, whilst little known or used, are beneficial where there is repetition of work either in number of units such as housing, or in repetitive floors in constructing or refurbishing multi-storey buildings.

The unit or floor numbers are shown on the vertical y-axis, and the timeline of the project on the horizontal x-axis.

Advantages of elemental trend analysis charts

- They are very suitable for repetitive work and strict trade sequencing as the flow of work can be clearly demonstrated.
- They are more sophisticated than bar charts and simpler than network diagrams.
- They clearly show the rate of progress of each trade.

Disadvantages of elemental trend analysis charts

- They are totally unsuitable for non-repetitive projects and largely unsuitable for limited repetition.

- They are not easily understood, user-friendly or readily understood by all practitioners.

(iii) Project network techniques (precedence diagrams, network analysis, critical path analysis)

Project network techniques comprise a number of methods but all are related to the dependency and precedence of one activity over another.

Networks may be represented as arrow diagrams where activities are represented by arrows and events (start and finishes) by 'nodes'. Activities take time and are represented as months, weeks, days or even hours, but events do not.

As stated earlier, the bar chart is a clear and simple method of plotting activities, but it has a clear disadvantage in that the relationship between activities cannot be properly represented. With networks, the relationship is immediately apparent and when a delay occurs, specifically affected activities will stand out.

A project network illustrates the relationships between activities (or tasks) in the project, showing the activities either on arrows or nodes.

Preparing network diagrams

The first operation, as with any method, is to compile a list of activities to be carried out.

One must then consider against each activity:

- Which activities must be completed before this activity can start?
- Which activities cannot start until this one is completed?

N.B.: One should always consider when an activity can be done rather than when one thinks it will be done.

Advantages of network diagrams

- They are extremely effective when one is considering highly critical activities with strict dependency of several activities upon each other.
- Float and critical path can be calculated fairly easily from such diagrams.

Disadvantages of network diagrams

- They require a great deal of time and effort to produce.
- They are not practical for 'normal' construction activities, which do not require highly detailed planning and monitoring of progress.
- They do not provide a ready format for financial reporting.

Most major projects are managed using software which combines high-quality presentation with ease of revision and updating. This software can also link the master programme to short-term programmes and also resource schedules and cash flow forecasts; however one must never forget the power of pen and paper.

3.6 The traditional approach of construction contracts

The value of a fully comprehensive construction programme is often misunderstood and undervalued by construction practitioners. This is often exacerbated by the minimal express requirement for a programme in most standard forms of construction contract.

Various commonly used forms merely state that the Contractor 'submits his master programme for the execution of the works', 'states the order in which he proposes to carry out the works' or 'shall submit a detailed time programme . . . and shall submit a revised programme whenever the previous programme is inconsistent with actual progress or with the Contractor's obligations', often without going into any detail as to the required content and structure of the programme, the timing of the submission, the timing of the submission of any revised programme or any formal acceptance of the Contractor's programme.

The danger is that, by understating the requirement for a programme within the contract, the true value of a programme is then misunderstood by the parties to that contract, and the programme submission and any acceptance can then descend into a 'box-ticking' exercise.

Clearly, although the programme requirement within these contracts is not spelt out in detail, it is advisable on any project for the Employer to require the Contractor to submit a programme, even though there may be no contractual requirement to do so, and it is also advisable for the Contractor to submit a detailed programme even if there is no contractual requirement to do so. The submitted programme should also identify the requirements to be met by the Employer.

The programme must be a dynamic tool which addresses all aspects of the project, including design, information and other requirements, and the procurement process and lead-in times for Subcontractors and Suppliers in addition to the construction operations. It must be flexible and reviewed and, if necessary, revised on a regular basis as the work progresses and various events and changes come to light, including the adjustments to float and time risk allowances.

3.7 The NEC approach

The NEC takes a different approach to other contracts in dealing with the programme. The ECC clearly states in detail what the Contractor is required to include in the programme he submits, with sanctions if he fails to do so, and also requires that the Project Manager either accepts the programme or gives his reasons for not accepting it.

Some practitioners say that the requirement within the ECC is very onerous and in some cases excessive. However, it is critical that the programme shows in a clear and transparent fashion what the Contractor is planning to do, when and how long each operation will last and what he needs for the Employer and others to be able to comply with it.

There is nothing within the programme requirements of ECC that a competent Contractor would not ordinarily include within a professionally constructed programme submitted for the benefit of himself and the receiving party. Whether and how he chooses to show the information could be another matter!

Such is the importance of a programme within the ECC that, if no programme is identified in Contract Data Part 2, one-quarter of the Price for Work Done to Date is retained until the Contractor has submitted the first programme showing the information the contract requires. The clause relates to the Contractor's failure to submit a programme showing the information the contract requires, not the acceptance of it (see Figure 3.3).

Contract:	CONTRACTOR SUBMISSION
Contract No:	C/S No..
Section A:	Submission of:
To: Employer	☐ Drawings ☐ Programme ☐ Test Results ☐ Other
The following information is transmitted for your acceptance:-	

Description	Date	No./Rev.	Copies

Response reqd. by ....
Signed: (Contractor) ....   Date ....

Section B: Reply
To: Contractor
The submission is returned:-

☐ Accepted   ☐ Accepted as noted
☐ To revise and resubmit   ☐ Rejected as noted

Notes:

Signed: (Employer) ....   Date:................

Copied to:

Contractor ☐ Employer ☐ File ☐ Other .... ☐

Figure 3.3 Typical template for submission of a programme and other documents.

The clear programme requirement enables it to be used to assess:

- the Contractor's ability to carry out the project in terms of the intended resources and within the contractual time frame;
- whether all of the contract needs and requirements have been considered;
- the Contractor's actual progress against that planned;
- the effect of compensation events upon Completion and Price.

Within the ECSC there is no specific requirement for a programme in any specific format, but Section 5 allows the Employer to identify whether a programme is required, what form it is to be in, what information is to be shown and when it is to be updated.

This requirement may vary from a simple list of dates to a fully resourced bar chart and in fact many practitioners 'copy and paste' requirements from the ECC when using the ECSC.

3.8 Dates in the contract

There are a number of dates in the ECSC which require some definition.

(i) *Starting Date*

In contrast with most forms of contract, the period of time for Completion of the Works is not stated. Instead, the Starting Date is decided and defined by the Employer in the Contract Data, together with the Completion Date. This has the advantage that tenderers know exactly when the works are required to start and to complete.

The Starting Date is stated in the Contract Data. The Contractor can start work from the Starting Date, and as there is no Access Date in the contract, assumes that he has the required access to the Site from the Starting Date. He also carries risks from the Starting Date and so has to provide the insurances in the contract prior to that date.

(ii) *Access Date*

There is no 'Access Date' expressly stated within the contract from which the Contractor may have access to and use of various parts of the Site. The Employer is merely obliged to allow access and use of the site to the Contractor as necessary for the work included in the contract (Clause 15.1).

It may be appropriate to include Access Dates within the contract, certainly if parts of the site are made available at different times within the contract.

Whilst the Contractor can start work on the contract from the Starting Date, he cannot start work on the Site until the first Access Date. If the Employer has stated different Access Dates for different parts of the Site, then the Contractor must programme his activities according to the dates when he will be given access.

(iii) Completion Date

The Contractor is obliged to complete the works on or before the Completion Date as stated in the Contract Data or as may be revised in accordance with the contract.

Whilst the work is progressing, the Contractor must submit a forecast of the date of Completion to the Employer each week from the starting date until Completion. This is an important obligation of the Contractor. Its purpose is to keep the Employer informed of the Contractor's likely date of Completion.

The Employer decides the date of Completion, and certifies within one week of Completion (Clause 30.3) (Figure 3.4). Note that the Employer is not deciding when the Contractor has to be completed by, but when he has completed. Failure to complete on time constitutes a breach of the Contractor's obligation.

If delay damages are included, the Employer is then entitled to pre-defined damages for each day of any delay.

Whilst many contracts use the terms 'Practical Completion' and 'Substantial Completion', the ECSC does not, as these terms are often subject to various interpretations.

The ECSC is more objective in defining Completion under Clause 11.2(1) as when the Contractor has 'completed the works in accordance with the Works Information except for correcting notified Defects which do not prevent the Employer from using the works and others from doing their work'.

The Contractor is required to comply with the requirements of the Works Information. This requirement may include not only physical Completion of the works, but also submission of 'as-built' drawings, maintenance manuals, training requirements and successful testing requirements. By including these requirements within the Works Information, Completion has not been achieved until they are completed.

Also, note the words 'except for correcting notified Defects which do not prevent the Employer from using the works and others from doing their work'.

The works may contain Defects, though these are not significant enough to prevent the Employer from practically and safely using the works and from others doing their work (Clause 11.2(1)).

This broadly aligns with the traditional terms 'Practical Completion' and 'Substantial Completion', where the Employer is able to take beneficial occupation of the works and use them as intended. Completion is certified with the requirement that the Contractor corrects the Defect before the end of the defect correction period following Completion.

The Employer is responsible for certifying Completion, as defined in Clause 11.2(1), within one week of Completion. Normally, the Contractor will request the certificate as soon as he considers he is entitled to it, but such a request is not essential.

```
┌─────────────────────────────────┬───────────────────────────────────┐
│ Contract: .........................    │   COMPLETION CERTIFICATE          │
│                                 │                                   │
│ Contract No: ....................  │   C/C No...........................   │
├─────────────────────────────────┴───────────────────────────────────┤
│                                                                       │
│ Completion achieved on: ......................... (Date)              │
│                                                                       │
│                                                                       │
│ The Completion Date is:  ........................... (Date)           │
│                                                                       │
│                                                                       │
│ The *defects date* is:      ......................... (Date)          │
│                                                                       │
│                                                                       │
│ The *defects on the attached schedule are to be*                     │
│ *corrected within the defect correction period*                      │
│ *which ends on:* .......................... (Date)                   │
│                                                                       │
│                                                                       │
│ The *outstanding works on the attached schedule are to be*           │
│ *completed by:* .......................... (Date)                    │
│                                                                       │
│                                                                       │
│ Comments:                                                             │
│                                                                       │
│                                                                       │
│                                                                       │
│                                                                       │
│                                                                       │
│                                                                       │
│                                                                       │
│ Signed: ...................................(Employer)     Date: ............... │
└───────────────────────────────────────────────────────────────────────┘
```

Copied to:

ContModel [ ] Employer [ ] File [ ] Other .... [ ]

Figure 3.4 Typical completion certificate.

3.9 Record keeping

One of the most important, but often understated, elements of managing a construction contract and its progress and any difficulties, whether one is a Contractor or an Employer, is the keeping of concise and timely records. It is beneficial if the records are agreed by the Contractor and the Employer. These are then undisputed records of fact. It is then only the interpretation of these facts which might subsequently be at issue.

A good record-keeping system will enable the Contractor or the Employer to notify early warnings promptly, assist the preparation of programmes, enable payments to be made and also prove invaluable in pricing and assessing compensation events, all of which have time scales applied to them, so readily available records are a must.

Records can include:

- Site diaries;
- Site meeting minutes;
- photographs, including dates;
- labour returns;
- Equipment returns;
- weather records;
- correspondence;
- financial accounts and records.

3.10 Instructions to stop or not to start work

The Employer has the authority to give instructions to the Contractor to stop or not to start work for any reason and may later instruct him to re-start it (Clause 30.4).

The instruction would be a compensation event (Clause 60.1(4)). However, if the instruction was required due to a fault of the Contractor, for example unsafe working, the Employer should notify the Contractor that the Prices and the Completion Date are not to be changed (Clause 61.2).

3.11 Acceleration

There is no provision for acceleration within the ECSC as there is with the ECC, though a number of Employers have included additional conditions to provide for it.

Acceleration in many contracts normally means increasing resources, working faster or working longer hours so that Completion can be achieved by the Completion Date; in effect the Contractor is catching up to recover a delay. However, within the ECC acceleration means bringing increasing resources, working faster or working longer hours so that Completion can be achieved before the Completion Date.

Under the ECC, the Project Manager may instruct the Contractor to submit a quotation for acceleration. He also states any changes to the Completion Date to be included within the quotation.

The quotation must include proposed changes to the Prices, and a programme showing the earlier Completion Date and Key Date. As with quotations for compensation events, the Contractor must submit details within his quotation; however the contract does not prescribe the basis of how the quotation is to be priced.

4 Defects

4.1 Introduction

Defects and the notification and correction of Defects are covered within the Engineering and Construction Short Contract (ECSC) by Clauses 40.1 to 43.1.

Within the Engineering and Construction Contract (ECC), there is a Supervisor appointed by the Employer, with most of the actions under Section 4 being between the Contractor and the Supervisor. The Contractor and Supervisor notify tests as and when required and carry out the various inspections as required by the contract; however the interactions within the ECSC are carried out between the Contractor and the Employer.

Any testing required must be clearly defined by the Employer within the Works Information, but the Contractor must also comply with any statutory requirements for quality and testing whether or not covered within the Works Information. Any materials, facilities and samples for testing, including who provides them, are also stated within the Works Information.

4.2 Quality management and control

Introduction

Quality management is a major management function within the construction industry and its various companies.

Unless a construction company can guarantee its Clients a quality product, it cannot compete effectively in the modern construction market.

Quality often stands alongside cost as a major factor in Contractor selection by Clients, as well as determining the efficiency of processes that the contractor adopts for Site operations. To be competitive and to sustain good business prospects, these quality systems need to be efficient and evidential.

The role of quality management for a construction company is not an isolated activity, but intertwined with all the operational and managerial processes of that company.

The modern concept of quality is considered to have evolved through three major stages over many years.

These stages are as follows:

(i)   *Quality control and inspection*

Inspection is the process of checking that what is produced is what is required. It is about the identification and early correction of Defects.

Quality control introduces inspection to stages in the Works, ensuring that it is undertaken to specified requirements. Quality control is done on a sampling basis dictated by statistical methods, i.e. sampling concrete cubes, mock-ups. The quality is measured by comparing the work actually carried out against the sample initially provided.

(ii)   *Quality assurance*

Quality assurance is the process of ensuring that standards are consistently met, thereby *preventing* Defects from occurring in the first place.

'Fit for purpose' and 'right first time, every time' are the principles of quality assurance.

(iii)   *Total quality management*

This is based on the philosophy of continuously improving goods or services. The key factor is that everyone in the company should be involved and committed from the top to the bottom of the organisation.

The successful total quality managed company ensures that their goods and services can meet the following criteria:

- be fit for purpose on a consistently reliable basis;
- delight the customer with the service which accompanies the supply of goods;
- supply a quality product that is so much better than the competition that customers want it regardless of cost. It has to be said that, whilst the construction industry prides itself on the ability to deliver quality projects, it is virtually unknown for Clients to disregard cost!

The pursuit of total quality is seen as a never-ending journey of continuous improvement.

The earliest and most basic form of quality management is quality control. This is described under the headings of:

1. quality control;
2. controlling quality;
3. quality control implemented in construction.

Quality control

The term 'quality control' is defined by an interpretation of its elements: 'quality' and 'control'.

Quality

The term 'quality' is often used to describe prestige products such as expensive jewellery and motor cars. However, although applicable to these items, the term 'quality' does not necessarily refer to prestigious products but merely to the fitness of the product to the customer's requirements. Quality is therefore described as meeting the requirements of the customer.

Control

The concept of being 'in control', or having something 'under control', is readily understood – we mean we know what we intend to happen, and are confident that we can ensure that it does.

Quality control is primarily concerned with defect detection and correction. The main quality control technique is that of inspection and statistical quality control techniques (i.e. sampling) to ensure that the work produced and the materials used are within the tolerances specified. Some of these limits are left to the inspector's judgement and this can be a source of difficulty.

The major objectives of quality control can be defined as follows:

- to ensure the completed work meets the specification;
- to reduce customers' or Clients' complaints;
- to improve the reliability of products or work produced;
- to increase customers' or Clients' confidence;
- to reduce production costs.

Controlling quality

The central feature to all quality control systems is that of inspection. To be effective the construction process requires that work items to be inspected must be catalogued into a quality schedule.

In the case of construction, inspection takes two forms: that which is quantifiable, e.g. line, levels, verticality and dimensions; and that which is open to the inspector's interpretation, e.g. cleanliness, fit, tolerances and visual checks.

There are some precise quantified inspections, including the commissioning of plant and machinery, pressure tests on pipe work and strength tests on materials such as concrete.

Statistical methods

These methods of quality control are based on the need to sample. In many of the processes of manufacture and construction the scale of the operation is too large to have 100 per cent inspection and therefore sampling techniques are employed.

The main techniques in statistical quality control are as follows:

- *Acceptance sampling,* based on probability theory, allows the work to continue if the items sampled are within pre-determined limits.
- *Control charts* compare the results of the items sampled with the result expected from a 'normal' situation. Usually the results are plotted on control charts which indicate the control limits.

In construction, it is the quality of materials that is normally controlled by statistical methods, the most common being that of the cube strength of concrete.

Quality control implemented in construction

Traditionally there are two sets of documents that are used to determine the required quality of a construction project. These are the specifications and the contract drawings. The Contractor uses these two documents during the Site operations stage of any project to facilitate 'quality' construction.

The process of actual construction is dissimilar to that of a production line in that there are no fixed physical and time boundaries to each operation of the process, hence the positioning and timing of quality inspections cannot be predetermined.

In construction, quality checks are undertaken as each operation or suboperation is completed.

The majority of quality checks are undertaken visually. Visual quality checks of each section of construction are undertaken by the Contractor's engineers and foremen and then by the resident engineers and inspectors to ensure it complies with the drawings and specification.

Quantifiable quality checks are also made during the construction stage. These include testing the strength of concrete cubes, checking alignment of brickwork and commissioning of services installations. The results of these quality checks are recorded and passed to the appropriate party.

Notwithstanding the existence of quality assurance and the emergence of total quality management, most Clients still engage inspectors to reassure themselves.

However, the impact and importance of the Client's inspectors are much reduced in a quality assured or total quality managed company.

Quality assurance

Whereas quality control focuses on Defect detection and correction during the Works, quality assurance is based on Defect prevention.

Quality assurance concentrates on the production or construction management methods and procedural approaches to ensure that quality is built into the production system.

Quality assurance is described under the following headings:

- evolution of quality assurance from quality control;
- definition of quality terms;
- quality assurance standards;
- developing and implementing a quality assurance system;
- quality assurance in construction.

4.3 Defining a Defect

A Defect is defined within the ECSC (Clause 11.2(3)) as:

- a part of the works which is not in accordance with the Works Information

The Works Information provides the reference point for what the Contractor has to do to provide the works. If the work done by the Contractor does not comply with the Works Information, then the Contractor is obliged to correct the Defect. In most cases the work done will fall short of the Works Information, but as the words 'not in accordance with' are used, work which exceeds the Works Information requirement would also be a Defect. Clearly, the Employer may wish, having discussed with the Contractor, to accept such a Defect!

A Defect may not necessarily mean that the work is not fit for purpose. The Defect could simply be a colour variation. For example, the Works Information stated a certain shade of red paint and the Contractor used a different shade of red paint, in which case a Defect exists as that part of the works was not in accordance with the Works Information. In such a case, unless the shade of red is a particular corporate colour or a stipulation of a regulatory authority, it may be prudent to accept the Defect.

It is also possible that work is 'defective' but as it complies with the Works Information, it is not a Defect! For example, the specification within the Works Information may stipulate use of a particular material which then fails. In that respect the work is defective but, as it complies with the Works Information, it is actually not a Defect.

4.4 Searching for and notifying Defects

Under Clause 40.1, the Employer has the authority to instruct the Contractor to search for a Defect. This action is normally defined as 'uncovering' or 'opening up' in other contracts. It is usually required in order to investigate whether a defect exists, and possibly the cause, and the correcting measures required.

Searching can include uncovering, dismantling, reassembly, providing materials and samples and additional tests which the Works Information did not originally require.

If the Employer instructs the Contractor to search for a Defect and no Defect is found, this is a compensation event (Clause 60.1(8)) and the Employer pays for the cost of the search and any delay and/or disruption as a result of that search. In addition if the search causes any delay to Completion the Contractor is awarded additional time to complete.

4.5 Correcting Defects

First, it is important to stress that the Contractor is responsible for meeting the quality standards required by the contract and has an obligation to correct Defects whether or not the Employer has notified him of them. In that respect he should adopt a quality assurance/quality control system (see above) which prevents Defects occurring and corrects them promptly if they do occur.

The Employer may notify a Defect to the Contractor at any time before the defects date (Figure 4.1).

If the Defect is notified:

- before Completion, the Contractor is obliged to correct the Defect before it would prevent the Employer or others doing their work;
- after Completion, the Contractor is obliged to correct the Defect before the end of the defect correction period, which begins at the later of Completion and when the Defect is notified.

Completion is defined under Clause 11.2.(1) as 'when the Contractor has completed the works in accordance with the Works Information except for correcting notified Defects which do not prevent the Employer from using the works and others from doing their work'.

The ECC has two periods for correction of Defects which are identified within the Contract Data:

1.   The period between Completion and the defects date for Defects which were not apparent or were not notified at Completion, or were notified, but did not prevent the Employer from using the works. This period is set within the Contract Data, and is normally 52 weeks. This is similar to the 'Defects Liability Period', 'Rectification Period' or 'Defects Notification Period' in other contracts, the period within which the Contractor is initially liable for correcting Defects.
2.   The 'defect correction period' is the period within which the Contractor must correct a notified Defect, failing which the Employer assesses the cost of having the Defects corrected by other people and the Contractor pays this amount. Note that, in this respect, although the Contractor has already paid the amount, the Employer may wish to leave the Defect uncorrected. In

Contract:	DEFECTS NOTIFICATION
Contract No:	D/N No..

The Contractor is notified of the following defects:

Description	Date Corrected

The defect correction period is:

Certified all defects
corrected .... (Employer)　　　　　Date: ....

Copied to:

Contractor　　　　　Employer　　　　　File　　　　　Other ....

Figure 4.1 Typical defects notification form.

practice many Employers will seek to have the Contractor remedy the Defects rather than bring in another contractor to do the work.

The defect correction period starts at Completion for Defects notified before Completion and when the Defect is notified for other Defects. The defect correction period is not effective prior to Completion; therefore if a Defect becomes apparent prior to this, the Contractor has an obligation to correct it in readiness for Completion (Figure 4.2).

Some Employers may wish to designate different defect correction periods for different Defects depending on their urgency.

A typical example of a defect correction period for a water company upgrading domestic water supplies is:

(i) Type A Defects – 4 hours

- major leaks;
- Defects related to health and safety.

(ii) Type B Defects – 24 hours

- minor leaks.

(iii) Type C Defects – 7 days

- cosmetic Defects;
- Defects which are not Defect Type A or B.

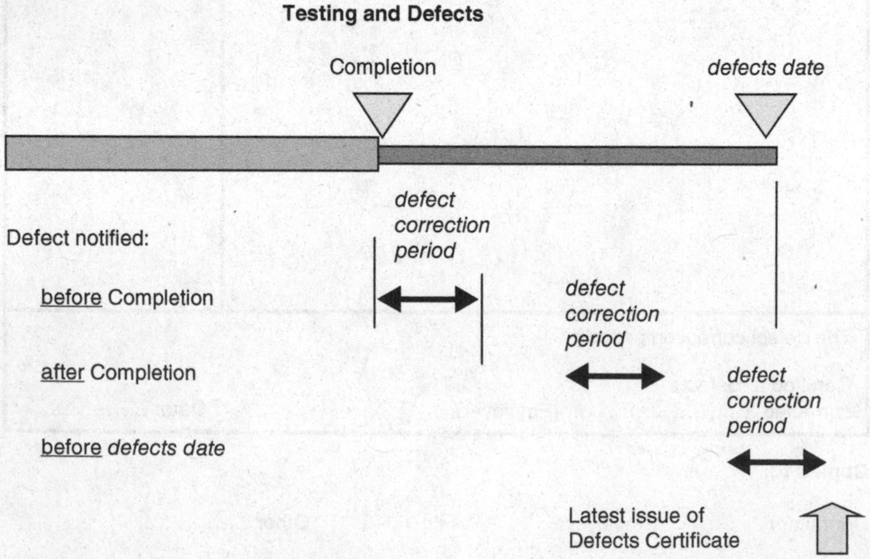

Testing and Defects

Figure 4.2 Testing and defects.

Defects which may prevent the Employer from using the work must be corrected before a Completion Certificate can be issued (Clause 11.2(2)).

4.6 Defects Certificate

The Defects Certificate is defined in Clause 11.2(4) as 'either a list of notified Defects which the Contractor has not corrected by the defects date or a statement that there are no such Defects'.

The Defects Certificate is issued by the Employer to the Contractor at the later of the defects date and the end of the last defect correction period.

Whilst many contracts require any Defects to be corrected before the Defects Certificate, or its equivalent is issued, under the ECSC the Defects Certificate is issued by the Employer at the appropriate time and may either list Defects which the Contractor has not corrected, or a statement that there are no outstanding Defects.

The Employer will normally only list what are usually defined as 'patent defects', i.e. those which are observable from reasonable inspection at the time, examples being a defective concrete finish or an incorrect paint colour, and may not include what are usually defined as 'latent defects', which may be hidden from reasonable inspection and may come to light at a later date, examples being some structural Defects.

The Contractor's and others' liability for correction of latent defects and other costs associated with them will be dependent on the applicable law, and liability may remain despite the issue of the Defects Certificate.

4.7 Repairs

Clause 43.1 requires the Contractor, until the Defects Certificate is issued and unless otherwise instructed by the Employer, to replace losses and repair damage to the works, Plant and Materials promptly regardless of how the damage has been caused.

4.8 The quality-related documents within the Works Information

Scope of Works

The Scope of Works must define clearly what is expected of the Contractor in Providing the Works. It must cover all aspects of the work, either specifically or by implication, otherwise it may be deemed to be excluded from the contract.

A comprehensive description should therefore be considered for the Scope of Works, which should be supplemented by specific detailed requirements in the form of drawings and specification. If reliance is placed solely on the Scope of Works, there is a danger that an item may be missed and could be the subject of later contention.

Whilst the Scope of Works must clearly describe the boundaries for the work to be undertaken by the Contractor, it may also describe the Employer's objectives and explain why the work is being undertaken and how it is intended to be used. It says what is to be done and what is to be included in general terms, but not how to do it.

The Scope of Works is the document that a tenderer can look to, to gain a broad understanding of the scale and complexity of the job and be able to judge his capacity as a Contractor to undertake it. It is written specifically for each contract and should be comprehensive, but it should be made clear that it is not intended to include all the detail, which will also be contained in the drawings, specifications and schedules.

Drawings

The drawings should detail all the work to be carried out by the Contractor and be complementary to the Scope of Works and the specification.

Specification

The specification is a written technical description of the character and quality of Materials and workmanship for the work which is to be carried out. It may also refer to the phasing or sequencing of the work to be carried out and also the outcomes to be achieved.

Note that there is no hierarchy or precedence between documents within the Works Information. If there is any ambiguity or inconsistency in or between the documents which form the Works Information, the Employer or Contractor should notify the other and the Works Information is changed by the Employer to correct the ambiguity or inconsistency. This is then a compensation event (Clause 60.1(1)) which is assessed as if the Prices and the Completion Date were for the interpretation most favourable to the Party which did not provide the Works Information (Clause 63.8).

Responsibilities

The responsibilities of the contracting parties and their representatives should be clearly defined in a manner which will leave no doubt as to the obligations that each is accepting. This will also allow the procedures necessary to enable the contract to progress satisfactorily from inception to Completion to be established at the outset of the work.

The contract as a whole will also define the responsibilities for the design, production and programming of the design information, the requirements for design approval, the supply of any free-issue materials, the issuing of instructions and the form which these instructions are to take, the programming of the works, the method of measuring and evaluating the work, the circumstances which will constitute a variation to the Scope of Works and the duties of the parties during construction and installation.

Testing take-over and liability for Defects

Various liabilities commence or are released when certain stages have been reached in the work. Such stages are usually acknowledged by the issue of a certificate, which may include certificates of handover and take-over, partial or final Completion and maintenance period.

Modern-day projects, by their very nature, will require tests to be conducted both during and on Completion of individual installations and in many cases in relation to a number of separate installations which operate in conjunction with each other to meet the design parameters of the project as a whole.

Testing therefore may relate to individual material components, factory testing of completed items of equipment off Site, pressure testing of individual sections of pipe work, testing of welds or pipe joints, a vast array of electrical and process control tests and tests of environmental conditions or quality of finished product after processing. Testing may also be required for insurance purposes.

Testing needs to be interpreted separately from pre-commissioning or commissioning, as does performance testing, where the whole or part of an installation requires certain preset specified design parameters to be met. The Employer may arrange for commissioning of completed installations or groups of installations to be carried out independently or jointly with the Contractor(s).

While the specification will set out the procedures and parameters for testing and commissioning, the contract will set out the obligations as to how these parameters are to be met and the consequences of failure to meet them. The responsibility for the provision of fuel for testing and testing equipment also needs to be stipulated. Performance testing is usually applicable where the design is the responsibility of the Contractor and can include the running of the completed Plant and the checking of product quality and quantity, feedstock consumed, use of energy, waste and by-products, environmental conditions and other aspects such as may be required. In such instances time limits for the rectification of performance Defects must be stated together with remedies for failure.

The documentation required in the form of test certificates, warranties and guarantees to enable proper records to be maintained should be stipulated, as should the effect in relation to the guarantee or warranty period and the extent to which it will apply.

In many cases the awarding of a certificate following testing will establish the date of take-over of the installation or, where applicable, that part ready for commissioning.

Liability for Defects should be defined in the contract together with the period of time for which such liability is to apply and the length of time within which Defects are to be remedied.

4.9 Limitation periods

Whilst the ECSC has a period from Completion to the defects date during which the Contractor is liable for correcting Defects, most countries also have a statutory limitation period for latent defects which come to light at a later date, beyond which a claim cannot be made for defective work.

In the UK, for example, the time limit for actions founded on simple contract is six years and for a specialty contract or deed, the time limit is twelve years. In both cases this is measured from the date on which the cause of action accrued; normally this is taken as the date of Completion.

4.10 Defining testing within the Works Information

Testing can be related to individual Plant and Materials on site, factory testing off site, testing of individual sections of work, welds and joints once they have been assembled or a multitude of mechanical and electrical tests.

Whilst the Works Information will set out the procedures and parameters for testing and commissioning, the contract will set out the parties' obligations as to how criteria are to be met and the consequences of failure to meet them.

Performance testing is usually applicable where the design of Plant and Materials is the responsibility of the Contractor. This can include the operation of the completed Plant and the checking of product quality and quantity, fuel consumed, use of energy, waste and by-products, environmental conditions and other aspects such as may be required. In such instances time limits for the rectification of performance Defects must be stated together with penalties for failure.

The documentation required in the form of test certificates, warranties and guarantees to enable proper records to be maintained should be stipulated, as will the effect in relation to the guarantee or warranty period and the extent to which it will apply.

In many cases the awarding of a handover certificate following testing will establish the date of take-over of the installation or, where applicable, that part ready for commissioning.

Liability for Defects should be defined in the contract together with the period of time for which such liability is to apply and the length of time within which Defects are to be remedied.

It is imperative that the Works Information, produced at tender stage and referred to in the Contract Data, defines the full extent and timing of tests and inspections required and the subsequent correction of any Defects.

This is very often not covered in sufficient detail within the Works Information, so it is important for the compiler of the Works Information to consider a number of key questions, as follows.

(i) What is the purpose of the test?

What is it intended to prove or disprove? Is the test required before a following piece of work can be carried out by the Contractor or another party, or is it required before the Employer can take over the works or a part of the works?

Contract: ……………………………..	EMPLOYER/CONTRACTOR NOTIFICATION
Contract No: …………………………	S/CN No…………………………………..

To: Contractor

Type of Notification/Instruction:	Date	Time
☐ Notification of test or inspection	…………	……………
☐ Notification of result of test or inspection	…………	……………
☐ Notification of identified Defect	…………	……………

Instruction to search:

Location and Details:

Result of test or inspection:
Test: PASSED/FAILED
Reason for failure:

It is planned to correct the Defect and retest on (date) ……………….

Defect corrected Yes/No   Date:………………….

Other comments or action:

Signed: (Employer/Contractor) …………………………………   Date: ……………

Copied to:

Contractor   Employer   File   Other ………………..

Figure 4.3 Typical Employer/Contractor notification.

(ii) What is to be tested?

What elements or components are required to be tested? Is it to be tested in an assembled or unassembled state?

(iii) How is the test to be carried out, and who is to provide the Materials, Facilities and Equipment to do the test?

Whether the test is a well-known 'industry standard' test or not, the exact nature of the test must be clearly defined. Who supplies the Equipment? Who pays for the power or fuel for the Equipment? Is a component, for example a concrete cube, to be tested to destruction? It is also important to detail who is to provide the Materials and Facilities to do the test and who meets the costs.

(iv) Who should do the test? Who should be present and who should be invited should they wish to attend when the test is taking place?

There may be a requirement for an independent party to carry out the test and provide results.

(v) When and where will the test take place? On site or off site?

Clearly state at what stage or on what date the test must take place. Sometimes tests may have to be carried out on a periodic basis, for example on concrete samples on delivery and as the concrete cures.

(vi) What is the expected result or outcome from the test?

What are we expecting the test to prove? What results do we need to gather once the test has been carried out?

(vii) Who should be advised of the outcome of the test?

The Employer will normally be advised of the outcome of a test, but sometimes a Subcontractor or a third party outside the contract may need to be informed.

(viii) What should be done in the event of a successful test, e.g. certification?

Does a certificate have to be issued following successful Completion of a test? Does that successful outcome mean that the Contractor can proceed to the next stage of the work? Is payment to the Contractor dependent on successful Completion of the test?

(ix) What should be done in the event of an unsuccessful test, e.g. rectification or replacement, and further testing?

Does the Contractor have to repeat the test? Is the retest done immediately after the first test, or should there be a period before the next test? Is it possible that even though something has failed a test, it may still be considered acceptable, possibly by the Contractor offering a cost saving?

(x) What about additional testing, i.e. to confirm a suspected Defect or the extent of a suspected Defect?

This may be referred to as a 'search' where the Contractor may be instructed to 'open up' work which is believed to be defective, possibly including the requirement for some form of additional testing.

There may be a requirement for Plant and Materials to be tested off Site before delivery.

The main documents within the Works Information in respect of testing and inspection will normally be the drawings and specification.

4.11 Accepting Defects

A Defect is defined as 'a part of the Works that is not in accordance with the Works Information' (Clause 11.2(3)).

In the event that a Defect becomes apparent, the Contractor corrects the Defect so that the part of the Works is in accordance with the Works Information.

The ECC includes provision for the Contractor and the Project Manager each to propose to the other that the Works Information should be changed so that a Defect does not have to be corrected. This may be included as an additional condition within the ECSC so that what is essentially a 'cosmetic' Defect can be accepted. Note that the latter part of Clause 60.1(1) implies a provision for accepting a Defect, but there is no express provision within the contract.

Box 4.1 Example

The Contractor has carried out decoration work; however the Employer has notified him that the wrong brand of paint has been used.

Whilst visually the quality of the work appears to be excellent, the work is essentially not in accordance with the Works Information and therefore a Defect exists.

The work is due to be completed by the end of the week, the decorating work being one of the final operations, so if the Contractor has to redecorate using the correct paint it will not only cost him money, but Completion may be delayed.

If both the Contractor and the Employer are prepared to consider changing the Works Information, then the Contractor could provide a quotation to the Employer stating a financial saving, based on him not having to correct the Defect or an earlier Completion Date or both. If the Employer accepts the quotation then the Works Information is changed and the Prices and/or Completion Date are changed in accordance with the quotation. However, note for the latter part of Clause 60.1(1) that the change to the Works Information does not create a compensation event.

---

5 Payment

5.1 Introduction

Payment, including assessment, certification and payment of amounts due, is covered by Clauses 50.1 to 51.2.

This section refers to two main terms:

(i) The Prices

These are the various elements that make up the total Price and are the amounts stated in the Price column of the Price List.

(ii) The Price for Work Done to Date

This term is used in making the assessment of amounts due to the Contractor, and is the total of:

- the Price for each lump sum item in the Price List which the Contractor has completed;
- where a quantity is stated for an item in the Price List, an amount calculated by multiplying the quantity which the Contractor has completed by the rate.

5.2 Assessing the amount due

The Contractor is required to assess the amount due to him, and by each assessment day, which is a date stated in the Contract Data he applies to the Employer for payment (Clause 50.1). There is an assessment day each month from the starting date until the month after the Defects Certificate has been issued (normally 12 months after Completion). Note that the Contractor's application for payment is a pre-condition to him being paid, so it is essential that he submits his detailed application on time (Figure 5.1).

There is no prescribed requirement for the application, but the Contractor must include sufficient details for the Employer to assess properly the amount due.

Payment

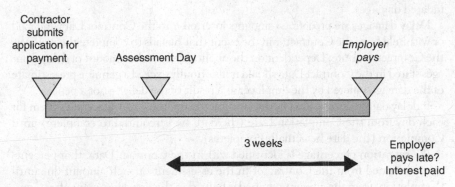

Figure 5.1 Application and payment timeline.

The Employer informs the Contractor of any amount the Contractor may have wrongly assessed within his application, prior to making the payment.

The amount due to the Contractor is:

- the Price for Work Done to Date,

which is made up of:

- the Price for each lump sum item in the Price List which the Contractor has completed, and
- where a quantity is stated for an item in the Price List, an amount calculated by multiplying the quantity which the Contractor has completed by the rate;

plus

- other amounts to be paid to the Contractor (including any tax which the law requires the Employer to pay to the Contractor).
  This will include VAT

less

- amounts to be paid by or retained from the Contractor.
  This will include retention and delay damages. Note that there is no retention-free amount as in the Engineering and Construction Contract (ECC).

Note that the Employer is not required to certify amounts due to be paid to the Contractor, though most commercial organisations using the Engineering and Construction Short Contract (ECSC) tend to issue written confirmation of acceptance of the Contractor's application, then the Contractor raises an invoice for the accepted amount.

Delay damages in the ECSC are normally referred to in other contracts as liquidated damages.

Delay damages are predefined amounts inserted into the Contract Data and paid or withheld from the Contractor in the event that he fails to complete the works by the Completion Date. Dependent on the applicable law, the amount of delay damages stated in the Contract Data should not normally exceed a genuine pre-estimate of the damage suffered by the Employer as a result of the delay – not a penalty.

If delay damages are stated in the Contract Data, the Contractor pays them for each day from the Completion Date (the date he is required to complete) until Completion (the date he actually completes).

If a retention percentage is identified within the Contract Data, that percentage is retained from the Contractor in the assessment of each amount due until Completion, then the amount retained is halved in the first assessment after Completion. No amount is retained after the Defects Certificate has been issued.

The ECSC retains the provision within the ECC with a sanction for failure to produce a first programme, should a programme be required (see Works Information).

It is essential that the Contractor either submits a first programme with his tender or within the time scale specified within the contract *if the Employer requires a programme to be submitted.* Any requirement for a programme, and its form and content, would be stated by the Employer within the Works Information.

Failure to do so will entitle the Employer to retain one-quarter of the Price for Work Done to Date in his assessment of the amount due. Note that the amount is only withheld if the Contractor has not submitted a programme which shows the information which the contract requires, e.g. method statement, provisions for float, if stated in the Contract Data. If the Contractor has submitted a programme which contains all the information that the contract requires, but the Employer disagrees with, for example, part of the method statement or the programme has not yet been accepted, then the provision does not apply.

This clause reflects the importance which attaches to the programme in managing works even of small value.

N.B.: In September 2011, the NEC3 publishers issued a brief amendment to the ECSC payment provisions to align it with the Local Democracy, Economic Development and Construction Act 2009:

Definitions

1.1

(1) The payment due date for an application for payment by the *Contractor* is the *assessment day* which follows receipt of that application.
(2) The final date for payment is three weeks after the payment due date.

Assessing the amount due

1.2 The *Contractor*'s application for payment is the notice of payment specifying the sum that the *Contractor* considers to be due at the payment due date (the notified sum). The *Contractor*'s application states the basis on which the amount is calculated and includes details of the calculation.

1.3 The following replaces subclause 50.4

If the *Employer* intends to pay less than the notified sum, he notifies the *Contractor* of the amount which the *Employer* considers to be due not later than seven days (the prescribed period) before the final date for payment. The *Employer*'s notification states the basis on which the amount is calculated and includes details of the calculation. A Party pays the notified sum unless he has notified his intention to pay less than the notified sum.

5.3 Payment

The Employer is required to make payment to the Contractor within three weeks of the next assessment day following receipt of an application from the Contractor. This clause has mistakenly been interpreted by some as the Contractor submitting an application by an assessment day, say 1 February, the payment not having to be made by the Employer until three weeks after the next assessment day, that next assessment day being 1 March. This is incorrect: the 'next assessment day' in this example would be 1 February.

Interest is paid by the Employer if a payment is late, or if it includes correction of an earlier payment. Interest is calculated from the date the correct payment should have been made until the date it is paid. The interest is calculated at the interest rate stated in the Contract Data, or if none is stated, at 0.5 per cent of the delayed amount per complete week of delay. This equates to 26 per cent per annum, which is very high, but it simplifies the interest calculation.

Box 5.1 Example

The interest due can be calculated for late payments on the basis of the following formula:

Payment due × interest rate × the number of complete weeks late

So if one assumes the following:

- payment due = £60,000
- payment 2½ weeks late
- no interest rate is stated in the Contract Data, therefore interest rate is 0.5 per cent

The calculation is:

£60,000 × 0.5% × 2 weeks

Interest due = £600.00

5.4 Project Bank Accounts

In 2008, the Office of Government Commerce (OGC) published a guide to fair payment practices, following which the NEC Panel prepared a document in June 2008 to allow users to implement these fair payment practices in NEC contracts.

This document, entitled Z3: *Project Bank Account,* can be introduced to the ECSC by means of a Z clause and authorises a Project Bank Account which receives payments from the Employer and is in turn used to make payments to the Contractor and Named Suppliers.

There is also a Trust Deed between the Employer, the Contractor and Named Suppliers containing the necessary provisions for administering the Project Bank Account. This is executed before the first assessment date.

The Contractor also includes in his subcontracts for Named Suppliers to become party to the Project Bank Account through a Trust Deed. The Contractor notifies the Named Suppliers of the details of the Project Bank Account and the arrangements for payment of amounts due under their contracts. The Named Suppliers will be named within the Contractor's tender, but also additional Named Suppliers may be included subject to the Project Manager's acceptance by means of a Joining Deed, which is executed by the Employer, the Contractor and the new Named Supplier. The new Named Supplier then becomes a party to the Trust Deed.

As the Project Bank Account is maintained by the Contractor, he pays any bank charges and also is entitled to any interest earned on the account. The Contractor is also required at tender stage to put forward his proposals for a suitable bank or other entity which can offer the arrangements required under the contract.

The process every month is that, at each assessment date, the Contractor submits an application for payment to the Employer, including details of amounts due to Named Suppliers in accordance with their contracts.

No later than one week before the final date for payment, the Employer makes payment to the Project Bank Account of the amount which is due to be paid to the Contractor. If the Project Bank Account has insufficient funds to make all the payments, particularly to the Named Suppliers, the Contractor is required to add funds to the account to make up the shortfall.

The Contractor then prepares the Authorisation, setting out the sums due to Named Suppliers. After signing the Authorisation, the Contractor submits it to the Employer for signature and submission to the Project Bank.

The Contractor and Named Suppliers then receive payment from the Project Bank Account of the sums set out in the Authorisation after the Project Bank Account receives payment.

In the event of termination, no further payments are made into the Project Bank Account.

6 Compensation events

6.1 Introduction

Compensation events, including their definition, notification, quotations, assessment and implementation, are covered within the Engineering and Construction Short Contract (ECSC) by Clauses 60.1(1) to 63.10.

When considering the principle of compensation events under the ECSC, it is appropriate first to consider how change is managed with most other standard forms of contract.

6.2 The 'traditional approach' to managing change

The 'traditional approach' to managing change in construction contracts is normally dealt with in three separate elements within the contract.

(i) Variation

The changes in the form of additions, omissions or substitutions are priced by measuring the changed work, then pricing the change using rates consistent with the Contractor's original tender. The pricing document for that tender is normally in the form of a bill of quantities, schedule of rates or a contract sum analysis.

Where the changed work is similar to that referred to in the pricing document, executed under similar conditions, and does not significantly change the quantity of work set out in the tender, the original rates and prices apply. Where the changed work is similar to that referred to in the tender, but not executed under similar conditions, and/or there is a significant change in the quantity of work, the original rates and Prices form the basis of the pricing, i.e. they are adjusted to allow for the differences in conditions and/or quantities. There are normally provisions for the parties to agree fair rates and Prices, consistent with the tender prices, and where the work cannot be clearly defined and/or measured then daywork provisions apply.

The problem is that there are rarely specific time scales within the contracts for submitting the Prices and agreeing them, apart from the Final Account, which is normally submitted by the Contractor and agreed by the Contract Administrator (Engineer, Architect, etc.) months or even years after the works have been completed.

(ii) Extension of time

Whilst the variation element deals with the changes to the Prices, the effect on the Completion Date is dealt with as a separate process, referred to as 'extension of time' or 'adjustment to Completion Date'. This provision allows for the Contract Administrator to extend time/fix a new Completion Date, preventing time becoming at large and thereby postponing the Employer's right to recover liquidated damages.

In that respect, the Contractor is required to submit a separate notice to the Contract Administrator upon it becoming apparent that a delay will occur. Normally there is a lengthy time scale within these provisions for the Contractor to submit his notice and in turn for the Contract Administrator to respond.

The purpose of the notice requirement is to allow the Contract Administrator to attempt to mitigate the problem either by stopping the delay from recurring or even omitting work, in addition to making a judgement on the Contractor's entitlement to additional time.

The notice will normally identify:

- the cause of the delay – detail of how and why the delay is occurring or likely to occur;
- the estimated effect on Completion – the Contractor should also give an estimate of delay in his notice so that the Contract Administrator can form his own opinion;
- the clauses on which the Contractor relies in requesting an extension of time.

(iii) Loss and expense

Finally, contracts have provision for the Contractor to recover loss and expense which he could not have recovered under any other provision within the contract.

Loss and expense claims have from experience been given a very bad name. In reality loss and expense claims should be a process by which the Contractor can submit a well-structured, fully substantiated claim to seek to recover any loss or expense which he has incurred as a result of something which is outside his control and/or foreseeability.

Whilst the financial effect of a variation should essentially be priced with the variation, the cumulative effect of a number of variations or other factors which are not the Contractor's risk may involve the Contractor in loss and expense in having to work uneconomically or out of sequence and this provision deals with that possibility. They are generally known as prolongation and disruption claims. Often agreement to loss and expense claims is a prolonged process which extends long after Completion of the project.

6.3 The NEC3 approach

The NEC3 contracts take a very different approach to pricing and managing change as contained within the compensation event provisions, whereby the event is notified, priced, assessed and implemented within a defined and relatively short time scale, including the cost and any time delay effect of the change.

Change Management

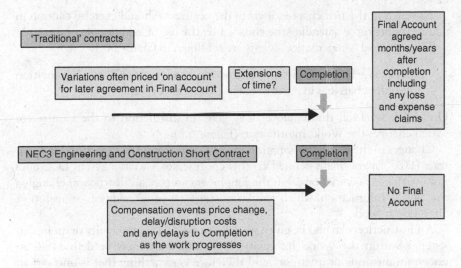

Figure 6.1 Comparison of change management provisions under 'traditional' contracts and the NEC3 Engineering and Construction Short Contract.

It recognises that it is critical that the cost and time effect of a change are considered together as the result of that change as a total cause and effect analysis. Therefore the Prices and the Completion Date are changed within a single provision rather than within three separate processes, as with the traditional approach previously described (Figure 6.1). The circumstances that constitute the basis of a compensation event are defined within the contract. This gives the contract drafter the facility to adjust the balance of risk between the Employer and Contractor.

6.4 Definition of a compensation event

There is no definition of a compensation event stated within the contract; however it could be defined as 'an event which is the Employer's risk, and if it occurs, and affects the Contractor, entitles the Contractor to be compensated for any effect on the Prices and any delay to the Completion Date'.

6.5 The compensation events

The compensation events may be found in a number of locations within the contract:

- fourteen events are stated in subclauses 60.1(1) to (14) – these are core clauses and therefore always apply unless amended within the contract;
- compensation events may be added, omitted or amended by the Employer in the Contract Data under the heading 'if additional conditions are required'.

These additional conditions are referred to in the other NEC3 contracts as 'Z clauses'.

As stated in the first chapter, users of the contracts should exercise caution in adding, omitting or amending the contract by the use of 'additional conditions'.

The principal compensation events are set down in Clause 60.1:

> 60.1(1) The Employer gives an instruction changing the Works Information unless the change is in order to make a Defect acceptable.

Under Clause 14.2, the Employer may give an instruction to the Contractor which changes the Works Information (Figure 6.2).

Changes to the Works Information in the Engineering and Construction Contract (ECC) are normally defined in other contracts as 'variations'. The ECC does not use the term 'variations', but the Employer can give an instruction changing the Works Information, which can be in the form of an addition, omission or alteration to work.

An instruction can also be given in order to resolve an ambiguity or an inconsistency within the Works Information. An ambiguity may be defined as 'an uncertain meaning or intention' and therefore is something that is unclear; an inconsistency may be defined as a 'conflict in meaning or intention' and therefore could be a mismatch or discrepancy between documents.

An example of an inconsistency is that the drawings in the Works Information provided by the Employer show a particular grade of concrete, but the specification shows a different grade. The Contractor or Employer should notify the other as soon as either becomes aware of the ambiguity or inconsistency and the Employer then gives an instruction resolving it. The subsequent change to the Works Information is then a compensation event under Clause 60.1(1).

Under Clause 63.8, a compensation event which arises from an instruction to change the Works Information to resolve an ambiguity or inconsistency is assessed 'as if the Prices and the Completion Date were for the interpretation most favourable to the party which did not provide the Works Information'. In this case it would be assessed in favour of the Contractor who is deemed to have priced the cheaper, easier or quicker alternative.

Note the reference within the clause to 'unless the change is in order to make a Defect acceptable'. This probably needs some explanation.

Box 6.1 Example

The Works Information states that the face of a concrete retaining wall should not deviate by more than 10 mm when measured with a 1 m straight edge. The Contractor has built the wall, but upon inspection the face of the wall deviates in some places by nearly 20 mm.

The definition of a Defect is 'a part of the works which is not in accordance with the Works Information'. Therefore a Defect exists and the Contractor is required to correct the Defect.

The Employer confirms that there is a Defect (it is important that the Parties recognise that there is a Defect). However, he is prepared to accept the wall as built; therefore the Employer gives an instruction changing the Works Information to state that the face of the wall should not deviate by more than 20mm. Thus the work is no longer defective and is therefore acceptable.

Contract:	EMPLOYER INSTRUCTION
Contract No:	EI No..
To: Contractor	☐ Instruction changing WI ☐ Proposal to change WI ☐ Other
Description	
Further instructions:	
☐ This instruction is/is not a compensation event	☐ Issued as proposed change. Do not implement.
☐ Provide quotation of time/cost effect	☐ Provide quotation of time/cost effect
☐ Start work immediately	☐ No cost instruction
☐ Keep records	
Signed: (Employer)	Date: ....
Contractor's Summary	
☐ Delay to completion is...............days/weeks ☐ No delay to completion ☐ Change in Defined Cost plus Fee Plus/Minus £ .... ☐ No change in Defined Cost plus Fee *Note: This is a summary only. Full supporting documentation or records must accompany this statement.*	
Signed: (Contractor)	Date:...............

Copied to:

Contractor   Employer   File   Other ....

Figure 6.2 Sample of Employer instruction.

Within the ECC there is specific provision for the Contractor and the Project Manager to propose to the other that the Works Information should be changed so that a Defect does not have to be corrected, in effect nullifying the Defect. This would not be a compensation event. This may be a specific consideration to add into the ECSC.

Also, within the ECC, a change to the Works Information provided by the Contractor for his design which is made either at his request or to comply with other Works Information provided by the Employer is not a compensation event. So Works Information provided by the Employer in the Contract Data takes precedence over that provided by the Contractor.

The wording in the ECSC is simpler within Clause 20.2 in that the Contractor does not start work which he has designed until the Employer has accepted that the design complies with the Works Information. So, again, Works Information provided by the Employer in the Contract Data takes precedence over that provided by the Contractor.

Box 6.2 Example

The Works Information included in the Contract Data states that all doors are to have brass furniture. The Contractor states in his tender that all doors will have aluminium furniture. The requirement for brass furniture in the Employer's Works Information in the Contract Data will prevail.

Thus it is critical that the Contractor ensures that, when he prepares and submits his tender, it complies with the requirements of the Employer's Works Information in the Contract Data.

60.(2) The Employer does not allow access to and use of the Site to the Contractor as necessary for the work included in this contract.

The Employer has a fundamental obligation to allow the Contractor access to the Site in accordance with the contract (Clause 15.1). Within the ECC, the Employer is obliged to allow access on or before the later of its access date and the date for access shown on the Contractor's Accepted Programme.

In that sense, if the Contractor shows a date in his programme when he requires access to and use of the Site or a part of the Site and the Project Manager accepts the programme, then the Employer is obliged to allow that access by the later of that date and the date in the Contract Data, failing which a compensation event occurs.

So the Contractor can relax the obligation to the date in the contract by showing a later date in the Accepted Programme, but cannot impose an earlier date than that stated in the contract.

60.1(3) The Employer does not provide something which he is to provide by the date for providing it stated in the contract.

The Employer is obliged to provide 'services and other things' as stated in the contract (Clause 15.2).

It is important that, in addition to the Employer stating the dates when he will provide things, the Contractor shows these requirements and the dates for providing them in his programme if he is required to submit one, so that the Employer is fully aware of what he needs to provide and the consequences of not providing it.

60.1(4) The Employer gives an instruction to stop or not to start any work.

It is important to note that the Employer giving an instruction to stop or not to start any work is always a compensation event, regardless of the reason why the instruction was given. However, if the Employer gives an instruction to the Contractor to stop a part of the work, for example because the Contractor is working unsafely and the Contractor notifies the Employer that he believes it is a compensation event, then under Clause 61.2, if the Employer decides that the event arose from a fault of the Contractor, he notifies the Contractor that the Prices and the Completion Date are not to be changed.

60.1(5) The Employer does not work within the conditions stated in the Works Information.

The words 'do not work within the conditions stated in the Works Information' can often cause debate. What work is the Employer required to do? Also, 'work within' implies a process over a period of time rather than a specific task or date, so it may be difficult to determine whether someone has worked within the conditions stated.

Again, it is important that, in addition to the Employer showing what he is going to do, the Contractor shows these requirements and the dates for providing them in his programme submitted for acceptance, so that the Employer is fully aware of what and when he needs to do something and the consequences of failing to provide it.

60.1(6) The Employer does not reply to a communication from the Contractor within the period required by this contract.

If the Employer does not reply to a communication, which could be, for example, the Contractor submitting design proposals, a programme or a quotation for a compensation event, then it is a compensation event as it is fundamentally a breach of contract.

However, if the Contractor notifies a compensation event in this respect, and the Employer decides that event has no effect upon Defined Cost or Completion, then again, under Clause 61.2, he notifies the Contractor that the Prices and the Completion Date are not to be changed.

The 'period required by this contract' may either be an expressly stated period of time or the more general 'period for reply' stated in the Contract Data.

60.1(7) The Employer changes a decision which he has previously communicated to the Contractor.

The ECSC refers to 'decisions' under the following clauses:

- Clause 30.3 – the Employer *decides* the date of Completion.
- Clause 61.2 – the Employer *decides* whether an event notified by the Contractor is a compensation event.
- Clause 61.3 – the Employer *decides* that the Contractor did not give an early warning of the event which an experienced Contractor could have given.
- Clause 61.4 – the Employer *decides* that the effects of a compensation event are too uncertain to be forecast reasonably.
- Clause 63.5 – the Employer has notified the Contractor of his *decision* that the Contractor did not give an early warning of the event which an experienced Contractor could have given.

Clause 60.1(7) has been interpreted by some NEC3 practitioners as a provision to allow the Employer to change and revise his assessment of an implemented compensation event. This is an incorrect interpretation. It must be stressed that the Employer does not have the authority to change his assessment once it has been made and notified to the Contractor. The Employer's assessment is final and the only recourse would be adjudication.

60.1(8) The Employer instructs the Contractor to search for a Defect and no Defect is found.

This relates to the Employer instructing the Contractor to search for a Defect under Clause 40.1.

Note that there is no provision for a 'mitigating factor' in that the search was only being instructed by the Employer because the Contractor gave insufficient notice of doing work obstructing a required test or inspection. If the search shows that there is no Defect, then it is a compensation event.

60.1(9) The Contractor encounters physical conditions which:

- are within the Site;
- are not weather conditions;
- an experienced contactor would have judged, at the date of the Contractor's offer, to have such a small chance of occurring that it would have been unreasonable to have allowed for them.

Only the difference between the physical conditions encountered and those for which it would have been reasonable to have allowed is taken into account in assessing a compensation event.

It is important to interpret this final paragraph correctly because if the Contractor did not allow anything in terms of Price and/or time within his tender for dealing with a physical condition, he should only be compensated for the difference between what he found and what he should have allowed, so the Contractor is unlikely to be compensated for the full value of dealing with the physical condition. If he should have allowed time and/or money within his tender for dealing with a physical condition then this must be considered in assessing a compensation event.

Note that, under Clause 60.2, in judging the physical conditions for a compensation event, the Contractor is assumed to have taken into account:

- *the Site Information* – this could include site investigations and borehole data;
- *publicly available information referred to in the Site Information* – this could include reference to public records about the Site;
- *information obtainable from a visual inspection of the Site* – note the use of the term 'visual'; the Contractor is not assumed to have carried out an intrusive investigation of the Site;
- *other information which an experienced Contractor could reasonably be expected to have or to obtain* – this is a fairly subjective criterion, but precludes the Contractor from relying solely on what is contained within the Site Information. Note that the clause refers to 'an experienced Contractor', not one who has a detailed local knowledge of the Site.

Note that there is no equivalent of NEC3 ECC Clause 60.3, 'if there is an ambiguity or inconsistency within the Site Information (including the information referred to in it), the Contractor is assumed to have taken into account the physical conditions more favourable to doing the work'. This could be the cheaper, easier or quicker alternative. As the clause is not included within the ECSC, each case must be decided on its merits, broadly relying on the rule of *contra proferentem* (*contra* = against, *proferens* = the one bringing forth) in that where a term or part of a contract is ambiguous or inconsistent it is construed strictly against the party which imposes or relies on it.

It is critical that the Employer makes all information in his possession available to the Contractor. He cannot be selective, withholding information with a view to obtaining advantage.

It is also important to note that if the Contractor encounters unforeseen physical conditions, which are often, but not always, ground conditions, he may not necessarily be compensated for the cost and time effect of dealing with them, so he must consider carefully the wording of Clauses 60.1(9) and 60.2 to prove his case.

It is also important to reiterate, as stated above, that compensation to the Contractor is assessed as the difference between what the Contractor found and what it would have been reasonable for him to have allowed in his tender, not simply the difference between what he found and what he *did* allow in his tender.

60.1(10) The Contractor is prevented by weather from carrying out all work on the site for periods of time, each at least one full working day, which are in total more than one seventh of the total number of days between the starting date and the Completion Date. In assessing this event, only the working days which exceed this limit and on which work is prevented by no other cause are taken into account.

This is a departure from the statistical basis in the ECC. It is much simpler and easier to implement. It has the effect that the Contractor takes the risk of adverse weather affecting the works up to a certain threshold, i.e. one-seventh of the total number of days in the contract period. Beyond that threshold if the Contractor is prevented from carrying out *all* work on the site and for *complete* working days by the weather and *no other cause*, then it is a compensation event. Adequate records will need to be kept to establish such a compensation event.

The problem with the wording of this clause, as distinct from the more comprehensive wording within Clause 60.1(13) of the ECC, is that it does not take into account 'normal' weather for that time of year at that location, e.g. it could be an exposed coastal site where the weather conditions are adverse for most of the time.

Box 6.3 Example

The Contractor is working on a project under the ECSC with a 12-week duration.

The total number of days between the starting date and the Completion Date is therefore $12 \times 7 = 84$ days. Therefore one-seventh of that total number is 12 days.

The Contractor bears the risk of being prevented from carrying out all work on the site for 12 full days. If he is prevented from carrying out all work on the site for any full days beyond that, delays and/or cost would be recovered as a compensation event.

60.1(11) The Employer notifies a correction to an assumption which he has stated about a compensation event.

Under Clause 61.4, if the Employer decides that the effects of a compensation event are too uncertain to be forecast reasonably, he states assumptions, the quotation being based on these assumptions (see example in Box 6.4).

If the Employer later notifies a correction to the assumption, it is a compensation event. In this case, the change to the Prices and any delay to Completion the Contractor had assumed in his original quotation are then replaced with corrected amounts.

Box 6.4 Example

The Employer instructs the Contractor to assume the cost of £2,000 for the people element of a quotation for a compensation event which is too uncertain to be forecast reasonably.

The cost of this element is subsequently recorded by the Contractor and found to be £2,600.

The Employer notifies a correction to the assumption, assuming a percentage for overheads and profit of 10 per cent. This is then priced as follows:

Cost from records	£2,600	
Less Employer's assumption	£2,000	
	£ 600	
Add percentage for overheads and profit @ 10%	£ 60	
Change to the Prices	£ 660	

60.1(12) An event which:

- stops the Contractor completing the works or
- stops the Contractor completing the works by the Completion Date

and which:

- neither Party could prevent;
- an experienced Contractor would have judged at the date of the Contractor's offer to have such a small chance of occurring that it would have been unreasonable for him to have allowed for it; and
- is not one of the other compensation events stated in this contract.

Clause 60.1(12) is a new 'no fault' clause introduced under NEC3, dealing with an event neither party could prevent, which stops the Contractor completing the works.

There has been much confusion and debate regarding these clauses since NEC3 was first published in 2005.

The intention of the drafters appears to be that this would be a *force majeure* (superior force) clause, often referred to as an 'Act of God' in other contracts and legal documents. 'Acts of God' include significant events such as flooding, earthquakes, volcanic eruptions and the associated dust clouds and other natural disasters which prevent the contracting parties from fulfilling their obligations under the contract. This clause essentially frees them, or gives some relief from their liabilities.

Whilst this appears to be what the drafters had intended, on closer examination the application of this clause is far wider.

If you take each element in turn:

Stops the Contractor completing the works

→ Due to the event, the works were never completed.

Stops the Contractor completing the works by the Completion Date

→ Due to the event, Completion of the works was delayed.

Neither party could prevent

→ Clearly, if either party could have prevented the event, then they would or should have. The clause refers to preventing the event, rather than taking some mitigating action to lessen its effect.

An experienced Contractor would have judged at the Contract Date to have such a small chance of occurring that it would have been unreasonable for him to have allowed for it

→ At tender stage, the likelihood of the event was so low, the Contractor would not have allowed for it.

Is not one of the other compensation events stated in this contract

→ If the matter could have been dealt with as one of the other compensation events, then it should be. Clause 60.1(12) then clearly deals with the exceptional event.

Examples of where this clause would at least cause confusion:

- Clause 60.1(10) has been deleted from the contract; one would then assume that the risk of unforeseen weather conditions is the Contractor's and he should have allowed for it.
- The Site is affected by a severe storm which occurs less frequently than once in 50 years, which delays Completion of the works by one week.
- The Contractor states that:

  (i) The weather stops him completing the works by Completion Date. Assume that the Contractor can prove that.
  (ii) Neither party could prevent the event. Whilst one can take measures to lessen the impact of a storm, one cannot prevent the storm from occurring.
  (iii) An experienced Contractor would have judged at the Contract Date to have such a small chance of occurring that it would have been unreasonable to have allowed for it. The Contractor states that he could not have predicted or allowed for such a rare event and its consequences.

(iv) It is not one of the other compensation events. If Clause 60.1(10) has been deleted from the contract, one could argue that it is no longer 'one of the other compensation events'.

Could Clause 60.1(12) also extend to loss or damage occasioned by insurable risks such as fire, lightning, explosion, storm, flood, earthquake and terrorism? Although the Contractor should insure against such risks and will not receive money from the Employer, the Contractor could still recover any delay to Completion through the compensation event.

If Clause 60.1(12) is intended as a form of *force majeure* provision, rather than including a fairly long-winded definition within the clause, why not just state: 'A *force majeure* event occurs' and allow the parties to interpret this internationally recognised term as they do in other contracts?

60.1(13) A difference between the final total quantity of work done and the quantity stated for an item in the Price List.

Any change (more or less) from the original quantities in the Price List to the final quantity is a compensation event.

This clause can create a compensation event very easily, as the final quantity of any item in the Price List only needs to vary by a minimal amount from the original quantity of the same item to create a compensation event, giving the Contractor the right not only to claim for the revised quantity (more or less) but also to claim additional time where appropriate.

60.1(14) A loss of or damage to the works, Plant and Materials which

- is not the fault or responsibility of the Contractor or
- could not have been prevented by any reasonable action of the Contractor.

The Contractor has a duty under Clause 43.1 promptly to make good any loss to the works, Plant and Materials and to replace any loss and repair any damage described. If such loss or damage is not the fault of the Contractor and could not have been prevented by any reasonable action by him (which could be a subjective issue), the Contractor is paid for the cost of the repair and any delay to Completion under the compensation event procedure.

N.B.: In September 2011, the NEC3 publishers issued a brief amendment to the ECSC adjudication provisions to align it with the Local Democracy, Economic Development and Construction Act 2009 and included the following additional compensation event: 'If the Contractor exercises his right under the Act to suspend performance, it is a compensation event.'

6.6 Comparison of compensation events under the ECSC and the ECC

Engineering and Construction Short Contract (ECSC)		Engineering and Construction Contract (ECC)	
60.1(1)	The Employer gives an instruction changing the Works Information unless the change is in order to make a Defect acceptable	60.1(1)	The Project Manager gives an instruction changing the Works Information
60.1(2)	The Employer does not allow access to and use of a part of the Site to the Contractor as necessary for the work included in this contract	60.1(2)	The Employer does not allow access to and use of a part of the Site by the later of its access date and the date shown on the Accepted Programme
60.1(3)	The Employer does not provide something which he is to provide by the date for providing it stated in this contract	60.1(3)	The Employer does not provide something which he is to provide by the date for providing it shown on the Accepted Programme
60.1(4)	The Employer gives an instruction to stop or not to start any work	60.1(4)	The Project Manager gives an instruction to stop or not to start any work or to change a Key Date
60.1(5)	The Employer does not work within the conditions stated in the Works Information	60.1(5)	The Employer or Others do not work within the times shown on the Accepted Programme, the conditions stated in the Works Information or carry out work on the Site that is not stated in the Works Information
60.1(6)	The Employer does not reply to a communication from the Contractor within the period required by this contract	60.1(6)	The Project Manager or the Supervisor does not reply to a communication from the Contractor within the period required by this contract
60.1(7)	The Employer changes a decision which he has previously communicated to the Contractor	60.1(8)	The Project Manager or the Supervisor changes a decision which he has previously communicated to the Contractor
60.1(8)	The Employer instructs the Contractor to search for a Defect and no Defect is found	60.1(10)	The Supervisor instructs the Contractor to search for a Defect and no Defect is found.........
60.1(9)	The Contractor encounters physical conditions...............	60.1(12)	The Contractor encounters physical conditions...............
60.1(10)	The Contractor is prevented by weather........	60.1(13)	A weather measurement is recorded............

(continued)

(continued)

60.1(11)	The Employer notifies a correction to an assumption which he has stated about a compensation event	60.1(17)	The Project Manager notifies a correction to an assumption which he has stated about a compensation event
60.1(12)	An event which stops the Contractor completing the works………..	60.1(19)	An event which stops the Contractor completing the works………..
In the ECSC but not included within the ECC		**In the ECC but not included within the ECSC**	
60.1(13)	A difference between the final total quantity of work done and the quantity stated for an item in the Price List	60.1(7)	The Project Manager gives an instruction for dealing with an object of value
60.1(14)	A loss of or damage to the works, Plant and Materials…….	60.1(9)	The Project Manager withholds an acceptance (other than acceptance of a quotation for acceleration or for not correcting a Defect) for a reason not stated in the contract
		60.1(11)	A test or inspection done by the Supervisor causes unnecessary delay
		60.1(14)	An event which is an Employer's risk stated in this contract
		60.1(15)	The Project Manager certifies take over of a part of the Works before both Completion and the Completion Date
		60.1(16)	The Employer does not provide materials, facilities and samples for tests and inspections as stated in the Works Information
		60.1(18)	A breach of contract by the Employer which is not one of the other compensation events in this contract

Figure 6.3 Comparison of compensation events under the Engineering and Construction Short Contract (ECSC) and the Engineering and Construction Contract (ECC).

Note that, in comparison with the ECC, there is no provision within the ECSC for the following:

(i) Dealing with objects of value or of historical interest

While Clause 70.1 states that the Contractor has no title to objects of value or historical interest and is not to move such an object unless instructed to do so by the Employer, there is no compensation event to cover it. An instruction to move an object would not change the Works Information so it would not be a compensation event under Clause 60.1(1) either.

(ii) The Employer withholding acceptance

An example where it would be advisable to include this is Clause 20.2, where the Contractor cannot start work which has been designed until the Employer has accepted that the design complies with the Works Information. If the Employer withholds or delays that acceptance, the Contractor is not compensated in terms of delay to Completion or the change in Price.

(iii) Tests or inspections by the Employer causing unnecessary delay

The Employer may either physically test or inspect something himself, or may appoint others to do so. Delays in that process are not compensation events.

(iv) Events which are the Employer's risk

The ECC includes risks held by the Employer, including:

- risks relating to the Employer's use or occupation of the Site;
- risks relating to items supplied by the Employer to the Contractor until the Contractor has received them, or up to the point of handover to the Contractor;
- risks relating to loss or damage to the works, Plant and Materials caused by matters outside the control of the Parties;
- risks arising once the Employer has taken over completed work, except a defect that existed at take-over, an event which was not an Employer's risk or due to the activities of the Contractor on site after take-over;
- risks relating to loss or wear or damage to parts of the works taken over by the Employer and any Equipment, Plant and Materials retained on Site after termination;
- any other risks specifically referred to in the Contract Data.

(v) Certifying take-over of a part of the works before both Completion and the Completion Date

The ECSC does not provide for take-over of a part of the works.

(vi) The Employer not providing materials, facilities and samples for tests

Any failure to provide, or delay in providing, is not a compensation event.

(vii) A breach of contract by the Employer which is not one of the other compensation events in this contract

This would provide a 'catch-all' to cover any breaches not covered by the other compensation events. An example could be the Employer has not paid a certificate and the effect of the non-payment is that the Contractor has to delay or stop progress.

6.7 Notifying compensation events

Under Clause 61.1, compensation events are notified by the Contractor to the Employer. However, the second bullet implies that the Employer may do so, presumably where the event is an instruction by the Employer.

If the Contractor notifies, he must do so within eight weeks of becoming aware of the event. Otherwise he is not entitled to a change in the Prices or Completion Date, unless the event arises from an instruction by the Employer.

The intention of the clause is to compel the Contractor to notify compensation events promptly, otherwise any entitlement to additional time and money is lost. Note that the eight-week rule requires the Contractor to notify the compensation event to the Employer within eight weeks of becoming aware of the event, not just to have given early warning or mentioned the possibility of a compensation event in a discussion with the Employer.

The clause therefore covers compensation events initiated by the Contractor, rather than the Employer, examples being the finding of unforeseen physical conditions or the weather and its impact, which the Employer may not have been aware of unless the Contractor had notified it.

An example of the Employer failing to notify a compensation event when he should have would be if an instruction was issued by the Employer to the Contractor changing the Works Information, but at the time the Employer did not notify the compensation event, and the Contractor in turn did not notify it either. Even if the compensation event is not notified within eight weeks, the responsibility remains with the Employer as he gave the instruction and should have notified the event to the Contractor but did not.

Various opinions have been published about the enforceability and effectiveness of time bars in contracts such as NEC3. In particular, commentators have debated whether the clause is a condition precedent to the Contractor being able to recover time and money, and whether a party, the Employer, can benefit from its own breach of contract to the detriment of the injured party, the Contractor.

For example, if the Employer does not provide something which he is to provide by the date for providing it stated in the contract, this is a valid compensation event under Clause 60.1(3). However, can the Employer prevent the Contractor from receiving any remedy because the Contractor diligently warned him several times that he was still awaiting information from him, but failed to notify a compensation event formally to the Employer within the eight-week limit, and possibly, if Completion is delayed, the Contractor would have to pay delay damages?

The author, whilst not being a lawyer, is of the opinion that the parties are clear as to what the terms of their agreement are at the time the contract is formed. It is also clear what happens if the Contractor does not notify a compensation event within the time stated within Clause 61.1. He is protected against notifications which the Employer should have given but did not, and therefore the time bar must be effective and enforceable.

N.B.: A compensation event is not notified after the defects date.

Clause 61.2 covers two possible outcomes to the Contractor's notification of a compensation event.

(i) Negative reply

If the Employer responds by stating that an event notified by the Contractor:

- arises from a fault of the Contractor;
- has not happened and is not expected to happen;
- has no effect upon Defined Cost or upon Completion; or
- is not one of the compensation events stated in the contract

he notifies the Contractor of his decision that the Prices and the Completion Date are not to be changed. The Employer only needs to name one of these as his reason for refusing a compensation event.

(ii) Positive reply

If the Employer decides otherwise, he notifies the Contractor within one week of the Contractor's notification that it is a compensation event, and instructs him to submit a quotation.

Under the ECC, if the Project Manager does not reply within one week of the Contractor's notification, or a longer period to which the Contractor has agreed, then the Contractor may notify the Project Manager to that effect. If the Project Manager does not reply within two weeks of the Contractor's notification, then it is treated as acceptance that the event is a compensation event and an instruction to submit quotations.

This may be useful to insert as an additional condition within the ECSC to provide for the situation where the Employer gives no response to the Contractor's notification, though clearly the Employer would then be making provision for his own failure!

6.8 Employer's assumptions

When the Contractor is instructed to submit a quotation, there may be a part of the quotation which is too uncertain to be forecast reasonably. In this case, under Clause 61.4, the Employer should state what the Contractor should assume within his quotation, which could be a cost and/or time effect. Subsequently, when the effects are known or it is possible to forecast reasonably, the Employer notifies a correction to the assumption and it is dealt with as a correction under Clause 60.1(11).

The assumption effectively acts as a Provisional Sum, relieving the Contractor of any risk for the work covered by the assumption.

Note that the Employer must state the assumption: if the Contractor makes assumptions when pricing the compensation event, then they are not subsequently corrected if later found to be wrong.

6.9 Quotations for compensation events

It is important to recognise that a quotation for a compensation event includes proposed changes to the Prices and any delay to the Completion Date. The Contractor is required to provide details of his quotation when he submits it.

Box 6.5 Example

The Contractor is instructed by the Employer to excavate a trial pit approximately 1 m × 1 m × 500 mm to establish the nature of the substrata in a part of the Site, so that the design for a new foundation can be finalised. The Employer tells the Contractor that he and the Structural Engineer will inspect the sides to the excavation as the Contractor progresses with the work and will instruct the Contractor when to stop excavating.

In this case the effects of the compensation event to excavate the trial pit are too uncertain, as the Contractor does not know the eventual size of the trial pit, so the Employer should state what the Contractor should assume, e.g. 1 m × 1 m × 500 mm, then notify a correction if the assumption is proven to be wrong.

An adjustment can then be made under Clause 60.1(11).

The quotation must be submitted within two weeks of the Contractor being instructed to do so by the Employer or, if no instruction is received, within two weeks of the notification of the compensation event by the Contractor or the Employer (Figure 6.4).

Under Clause 62.2 the Employer can instruct the Contractor to submit a quotation for a proposed instruction or changed decision. This is not an instruction to do work but to provide a quotation for something that is proposed. Once the Contractor submits his quotation, he must await further instruction to do the work.

A problem that has existed since the ECSC was originally launched in 1999 was, if the Contractor failed to provide a quotation for a compensation event, there was no resolution for the Employer. The April 2013 reprint with amendments has now included a new Clause 62.3, 'if the Contractor does not provide a quotation for a compensation event within the time allowed, the Employer assesses a compensation event and notifies the Contractor of his assessment'.

Once the Employer has instructed the Contractor to submit a quotation, the Contractor must submit his quotation to the Employer within two weeks. Note that, unlike the ECC, there is no provision for this period to be extended.

Submission and assessment of quotations

Contractor submits a quotation	Employer replies to quotation
2 weeks	2 weeks
No provision to extend the time for the Contractor to submit a quotation	No provision to extend the time for the Employer to reply to the quotation

Figure 6.4 Quotation times.

Once the Contractor has submitted his quotation to the Employer, the Employer is required to reply to the Contractor as follows.

For a proposed instruction:

- He could notify that the proposed instruction or changed decision will not be given.
- He could notify the instruction or changed decision as a compensation event and accept the quotation.
- He could notify the instruction or changed decision as a compensation event and that he does not agree with the quotation.

For other compensation events:

- He could notify acceptance of the quotation.
- He could notify that he does not agree with the quotation.

If the Employer does not agree with a quotation, the Contractor may submit a revised quotation within two weeks of the Employer's reply. If the Employer does not agree with the revised quotation, or if none is received, he can assess the quotation himself and notify the Contractor of his assessment.

If the Employer makes his own assessment, he should put himself in the position of the Contractor, giving a properly reasoned assessment of the effect of the compensation event, detailing the basis of his calculations and providing the Contractor with details of that assessment. An Employer's assessment is not simply the Contractor's quotation returned to the Contractor with 'red pen' reductions down to a figure the Employer is prepared to accept or to pay!

It is also important to recognise that the Employer is not sending his assessment to the Contractor for his acceptance. It is the final decision under the contract, and the Contractor's only remedy is adjudication.

There has always been concern amongst Contractors who were working under the earlier editions of the NEC who queried what remedy was available if there was a failure to make an assessment within the period required by the contract.

The simple answer is that failure to reply to a communication within the period required by the contract is another compensation event, in the case of the ECSC under Clause 60.1(6), which does not remedy the problem with the initial compensation event!

The drafters of the NEC3 contracts considered the problem and introduced a new clause. For example, under the ECC, under new Clause 64.4, if the Project Manager does not reply to a quotation within the time allowed, the Contractor may notify him to that effect. If the Project Manager does not reply within two weeks of the notification, the notification from the Contractor is treated as acceptance of the Contractor's quotation by the Project Manager. However, this new clause has not been included within the ECSC.

The Contractor cannot appeal against the assessment. If he does not agree he may refer the matter to adjudication. There is no contractual provision in NEC3 contracts to 'mutually agree' the Prices.

6.10 Delays to Completion

Note that, under Clause 62.1 within the Contractor's quotation he needs to consider and include any delay to the Completion Date. Clause 63.4 states that a delay to the Completion Date is assessed as 'the length of time that, due to the compensation event, Completion is forecast to be delayed'.

Note that the clause states 'any delay', not 'any effect', so the Completion Date can only be changed to a later date or remain unchanged due to a compensation event.

This is different wording to the ECC, which states 'a delay to the Completion Date is assessed as the length of time that planned Completion is later than planned Completion as shown on the Accepted Programme', so under the ECC any terminal float attached to the whole programme is retained by the Contractor. In contrast under the ECSC, the Contractor actually has to prove a delay, i.e. he cannot retain the terminal float.

From these provisions one can see that no compensation event can result in an earlier Completion Date, though it may give rise to an earlier planned Completion.

It is worth noting that the Contractor will usually show as little float as possible in his programme, i.e. all operations are close to or on the critical path.

6.11 Alternative quotations

Under Clause 62.6, the Employer may require the Contractor to provide alternative quotations, for example to carry out the work using existing resources or provide additional resources which might mitigate delay.

Discussion is required to take place between the Employer and the Contractor to ascertain what is feasible and practicable. This additional condition gives the Employer a choice of quicker/more expensive versus slower/less expensive options.

6.12 No reply to a quotation

There has always been concern amongst Contractors who were working under the earlier editions of the NEC who queried what remedy was available if the there was a failure to reply to their quotation within the period required by the contract.

The simple answer is that failure to reply to a communication within the period required by the contract is another compensation event, in the case of the ECSC under Clause 60.1(6), which does not remedy the problem with the initial compensation event!

The drafters of the NEC3 considered the problem and introduced a new clause.

For example, under the ECC, under new Clause 62.6, if the Project Manager does not reply to a quotation within the time allowed, the Contractor may notify him to that effect. If the Project Manager does not reply within two weeks of the notification, the notification from the Contractor is treated as acceptance of the Contractor's quotation by the Project Manager (Clause 62.6).

However, this new provision does not apply under the ECSC; therefore the only remedy is another compensation event under Clause 60.1(6).

6.13 Assessment of compensation events

Under the ECSC, there are two methods of pricing compensation events:

(i) For compensation events which only affect the quantities of work shown in the Price List

The change to the Prices is assessed by multiplying the changed quantities by the appropriate rates in the Price List.

(ii) For other compensation events

The change to the Prices is assessed by forecasting the effect of the compensation event upon the Defined Cost or if the compensation event has already occurred, due to the event that has occurred.

Effects are assessed separately for:

- people employed by the Contractor
  This relates to the cost of People who are directly employed by the Contractor, excluding Subcontractors, and must include management as well as operatives.
  The cost of design should also be considered, though if that is subcontracted it would be covered below.

- Plant and Materials
  Plant and Materials are 'items to be included in the works'.
  The cost of delivery, providing and removing packaging and unloading should be included. The cost of disposal of Plant and Materials should be credited.

- work subcontracted by the Contractor
  Subcontracted design should also be considered.

- Equipment
  Equipment is 'items provided by the Contractor, used by him to Provide the Works and not included in the works', so this is what other contracts refer to as Plant.
  This would include excavators, cranes, scaffold and temporary accommodation.
  The cost of transporting Equipment to and from the site and the erection and dismantling of the Equipment should be included.

Note that the amount for Equipment includes amounts paid for hired Equipment and an amount for the use of Equipment owned by the Contractor which is the amount the Contractor would have paid if the Equipment had been hired.

Defined Cost is defined in Clause 63.3 as open market or competitively tendered prices with deductions for all documents, rebates and taxes deducted.

In addition the following should be deducted from Defined Cost:

- the cost of events for which the contract requires the Contractor to insure;
- other costs paid to the Contractor by insurers.

The Contractor must ensure that he includes within his quotation cost and time which have a significant chance of occurring and are his risk (Clause 63.6). This should include adverse weather conditions and physical conditions which would not qualify as compensation events.

Also, the assessment of compensation events is based on the assumption that the Contractor reacts competently and promptly to the compensation event, with any additional cost and time due to the compensation event reasonably incurred. This provides some protection to the Employer against any inefficiency and incompetence of the Contractor.

There are two percentages for overheads and profit within the Contractor's Offer, which are applied to Defined Cost:

- the percentage for overheads and profit added to the Defined Cost for people;
- the percentage for overheads and profit added to other Defined Cost.

There is no schedule of items covered by the overheads and profit percentage which is added to Defined Cost. The following list, whilst not exhaustive, gives some examples of what would normally be considered as overheads and profit:

- the cost of off-site offices, e.g. the Contractor's head office;
- insurance premiums;
- performance bond costs;
- corporation tax;
- advertising and recruitment costs;
- sureties and guarantees required for the contract;
- some indirect payments to staff;
- profit.

The pricing of compensation events is much simpler than that in the ECC, which requires reference to the Schedule of Cost Components and the Shorter Schedule of Cost Components.

Note within the April 2013 amendment the additional Clause 63.6: 'the cost of preparing quotations for compensation events is not included in the assessment of compensation events.'

Clause 63.9 states that 'assessments for changed Prices for compensation events are in the form of changes to the Price List'.

Clause 63.10 emphasises the finality of the assessment of compensation events. If the subsequent records of resources on work actually carried out show that achieved Defined Cost and timing are different from the forecasts included in the accepted quotation or in the Employer's assessment, the assessment is not changed. The only circumstances in which a review is possible are those stated in Clause 61.4.

Box 6.6 Example

The Employer gives an instruction to the Contractor as follows:

Construct a 20 m long × 150 mm diameter drainage run, as shown on Drawing No. 36.

The Contractor's quotation will comprise changes to the Prices and also any delay to the Completion Date.

Delay to Completion

First, the Contractor considers whether there will be any delay to the Completion Date. The Contractor should provide details to demonstrate to the Employer the basis on which the delay has been calculated. This will probably require him to issue a programme or an extract from a programme with his quotation.

He assesses that this additional work will take him two weeks to complete and, as the instruction has been issued only three weeks before planned Completion, by the time he has obtained the materials to carry out the work it will delay the Completion Date by one week.

Delay to the Completion Date has therefore been assessed as one week.

Changes to the Prices

There is no drainage of this type within the Price List, so the Contractor must assess the changes to the Prices by forecasting the effect of the compensation event upon Defined Cost using the following components:

- people employed by the Contractor;
- Plant and Materials;
- work subcontracted by the Contractor;
- Equipment.

The two percentages for overheads and profit are then applied to Defined Cost (Figure 6.5).

Construct a 20 metres long × 150 mm diameter drainage run as shown on Drawing No. 36.		
PEOPLE		
Labourers		
2 labourers × 2 weeks × £350 per week	1400.00	
Ganger (part time)		
1 ganger × 2 weeks × 20% × £420 per week	168.00	
JCB Operator		
Note that the JCB Operator is priced with the "People" not the "Equipment".		
JCB Operator		
1 week x £500 per week	500.00	
As Completion is delayed, the Contractor must also consider people costs in respect of the extended time (Preliminaries Costs)		
Site Supervision		
Site Manager		
1 week × £675 per week	675.00	
Section Foreman		
1 week × £520 per week	520.00	
Total People Cost:		3263.00
Add percentage for overheads and profit in respect of Defined Cost for people	15%	489.45
PLANT AND MATERIALS		
22 metres × 150mm diameter pipe × 14.50 per metre	319.00	
(20 metres + 10% waste)		
Total Plant and Materials Cost:		319.00
EQUIPMENT		
Excavator × 1 week × £1150 per week	1150.00	
Fuel – 300 litres per week × 1.80 per litre (11 litres per hour when working)	540.00	

(continued)

(continued)

Again, as Completion is delayed, the Contractor must consider Equipment costs in respect of the extended time within the Working Areas		
Temporary Office Hire including furniture and other Equipment		
1 week × £180 per week	180.00	
Temporary Toilet Hire		
1 week × £100 per week	100.00	
Total Equipment Cost:	1970.00	1970.00
Add percentage for overheads and profit in respect of Defined Cost for Plant and Materials, and Equipment	12%	274.68
Total of Quotation:		**6316.13**

Figure 6.5 Worked example of a quotation for a compensation event.

6.14　Remedy for failure to give early warning

As stated previously in Chapter 1, under Clauses 61.3 and 63.5, if the Employer decides that the Contractor did not give an early warning of the event which an experienced Contractor could have given, he notifies this to the Contractor when he instructs him to submit a quotation. If the Employer has given this notification, the event is assessed as if the Contractor had given early warning.

6.15　Final Certificate

It is worth mentioning that the NEC3 contracts do not have the equivalent of a Final Account or Final Certificate which is found in other contracts, certifying that the contract has fully and finally been complied with, and that issues such as payments and compensation events have all been dealt with.

Some may say that, because of the compensation event provisions, there is no Final Account, and, because the Defects Certificate confirms correction of any outstanding Defects, there is no need for a Final Certificate. However a Final Certificate could give matters some closure.

6.16　Appendix to Chapter 6

Figures 6.6, 6.7 and 6.8 show suggested templates for notifying and managing compensation events.

Contract: Contract No:	COMPENSATION EVENT NOTIFICATION CEN No..
To: Contractor/Employer	

You are notified of the following compensation event:

Description

The Compensation Event is likely to have an effect on:
- ☐ Price
- ☐ Completion

Has an Early Warning Notice been given?
- ☐ Yes
- ☐ No

Quotation Required?
- ☐ Yes
- ☐ No

Signed: Employer/Contractor:.............................. Date: ....

Copied to:

Contractor   Employer   File   Other ....

Figure 6.6 Template for compensation event notification.

Contract: Contract No:	COMPENSATION EVENT QUOTATION CEN No...
To: Employer	
<u>Summary of Quotation</u>	
Change to Prices:	
Change to Completion Date:	
Signed: .... (Contractor) Date:	

The Employer's reply is:

Proposed instruction:

    ☐ A notification that a proposed instruction will not be given
    ☐ Notification of an instruction and acceptance of the quotation
    ☐ Notification of an instruction and non agreement of the quotation

Other compensation events:

    ☐ Acceptance of the quotation
    ☐ Notification of non agreement of the quotation

Signed: .... (Employer)   Date: ....

Copied to:

Contractor   Employer   File   Other ....

Figure 6.7 Template for compensation event quotation.

Contract:
....

Contract No:
....

REGISTER OF EARLY WARNINGS AND COMPENSATION EVENTS

Ref.	Date notified	Origin E/C	Description of event	Clauses	Ref.	Action by E/C	Date Concl.	Notes
1								
2								
3								
4								
5								
6								
7								
8								
9								
10								
11								
12								
13								
14								

Figure 6.8 Template for register of early warnings and compensation events.

7 Title

7.1 Introduction

Title, in the sense of ownership, is covered within the Engineering and Construction Short Contract (ECSC) by Clauses 70.1 and 70.2.

7.2 Objects of value

In the event that the Contractor discovers an object of value or of other interest, such as archaeological remains within the site, he has no title to it. The Contractor must also not move such an object: he notifies the Employer, who then instructs the Contractor how to deal with it. It is vital that the Contractor only takes instructions from the Employer in this respect, although particularly in the case of archaeological remains, other parties such as local authorities, museums and educational establishments may have an interest in the find and how it is dealt with, but instructions under the contract should still come from the Employer.

It is interesting to note that there is no direct compensation event as a result of such a finding as there is under the Engineering and Construction Contract (ECC) (Clause 60.1(7)). The ECSC Guidance Notes state that the Contractor should ask for an instruction from the Employer if something of value or historical interest is found, and such an instruction may be a compensation event under Clause 60.1(1); however, such an instruction would probably not change the Works Information, so it would not be a compensation event under Clause 60.1(1) unless it did change the Works Information.

7.3 Title to Materials from excavation or demolition

Unless otherwise stated in the Works Information, the Employer has the title to materials from excavation or demolition (Clause 70.2).

It is important in this respect that the Works Information clearly details any title the Contractor has to materials as he may wish to sell, recycle or remove those materials, and also if the Employer has title to the materials, he must clearly state what the Contractor is to do with them; for example, hand them to the Employer on site, transport them to another location.

7.4 Title to Plant and Materials on site

The general principle with title to Plant and Materials brought on to a construction site by the Contractor is that, in the absence of any contractual provision to the contrary, title will remain with the Contractor until they are finally incorporated (built) into the works, at which point they will become the property of the Employer. Once title has passed, the Contractor cannot, without an express right in the contract, normally with the consent of the relevant Contract Administrator, remove the Materials.

7.5 Title to Plant and Materials off site

A provision to consider which is not covered by the ECSC is the possibility of the Contractor requiring payment for Materials held off site awaiting delivery.

It may be appropriate to consider including the following additional conditions when considering payment for Plant and/or Materials off site, with the following to be confirmed in writing by the Contractor:

- The Plant/Materials are intended to be incorporated into the works.
- The Plant/Materials are in accordance with the contract.
- Nothing remains to be done to the Plant/Materials prior to being incorporated into the works; for example, a steel staircase should look like a steel staircase, not a pile of unidentifiable stock steel.
- The Plant/Materials have been set aside from other stock, and clearly labelled as destined for the specific project to which the contract relates.
- The Plant/Materials are vested in the Contractor, who then transfers title in accordance with the contract, free from any charges, liens or encumbrances.
- The Plant/Materials are insured against loss or damage for their full value under a policy of insurance protecting the interests of the Employer and the Contractor until they are delivered to the site.
- The Plant/Materials can be inspected at the storage location at any time.
- The Plant/Materials cannot be removed from the storage location, except for delivery to site.

There should also be provision for a representative of the Employer to visit the location of off-site Plant/Materials to evidence that they are being secured and protected in accordance with the contract.

The issue of title to Plant/Materials does not normally tend to cause problems unless a party becomes insolvent, when it then often becomes a major problem as a party such as an Employer, who believed he had good title to unfixed Plant/Materials because the contract stated that when they were on site he had title, finds that he does not have title.

Whilst this book is intended for international use rather than purely focusing on English law and case law, it is not unknown for an Employer to pay a Contractor for unfixed materials on site, but due to the insolvency of a Subcontractor and the Contractor not having title to those materials, the Employer then also has to pay the Subcontractor!

One aspect to bear in mind in considering unfixed materials on or off site is the use of 'retention of title' clauses.

A retention of title clause, which normally states 'title of the goods remains with the seller until the purchase Price is paid in full', is one that allows the Supplier of materials to retain ownership of those materials until they are fully and finally paid for, irrespective of any contractual provisions or agreements between, say, the Contractor and the Employer. Sometimes a retention of title clause may state 'title in the goods remains with the seller until the Price and all other sums owing by the purchaser to the seller are paid in full'.

Such clauses will override any clauses in the main contract that may stipulate that title will pass to the Employer when materials are delivered to the site, when they are marked or when the Employer has paid for them as the Supplier is not a party to the contract. These clauses then provide security for the Supplier against the purchaser's insolvency.

In order to be effective, retention of title clauses must be stipulated and accepted at the time the contract is formed, i.e. when the supplier's quotation is offered and accepted.

In order to provide this protection when insolvency occurs, the receiver or liquidator must be satisfied on each of the following points:

- The wording of the clause includes the specific goods for which protection is being claimed.
- The clause has been incorporated into the contract for the supply of those materials.
- Any goods claimed can conclusively be identified.
- The goods supplied relate to an unpaid invoice.

The goods being claimed under the retention of title clause must be identifiable and the receiver must be fully satisfied that any items claimed are the actual goods supplied by the Supplier claiming them. Most retention of title clauses relate to raw materials, stock in trade or livestock.

The best method of identification is where the goods are marked with the name of the Supplier or where their serial numbers are quoted on any unpaid invoices.

The Supplier should be allowed access to the insolvent party's premises to inspect the goods which he considers are subject to his claim. Of course, the Supplier should not be allowed to remove any goods until the official receiver or party acting for the insolvent party is satisfied that the claim is valid.

If a retention of title clause was drafted only to retain ownership of the goods until payment was made for them, goods such as any raw materials cease to be caught by it once the manufacturing process has begun, i.e. once the goods have lost their identity. For example, leather supplied in the production of handbags, once cut, is deemed to have created a new product.

It may be possible for the Supplier to retain title to the goods supplied even if they have been used in a manufacturing process provided they are still identifiable, in their original form and are easily removable. For instance, ceiling tiles supplied to be installed into a suspended ceiling do not lose their identity because they are fixed within the ceiling grid. However, they must be capable of being removed without damaging the ceiling grid.

8 Indemnity, insurance and liability

8.1 Introduction

Risks and insurance, including their definition, indemnity and cover requirements and policies, are covered within the Engineering and Construction Short Contract (ECSC) by Clauses 80.1 to 82.1.

Insurance is, in simple terms, the principle of many paying in for the few to be compensated. It is basically a contract between the 'insured', who offers to pay an agreed sum, called a premium, to the 'insurer', who warrants to pay out an agreed sum should a particular and specified event occur, e.g. death or damage to property. In effect, it is possible to insure against anything which holds an element of risk.

8.2 Types of insurance

The required insurances under a construction contract can normally be grouped under the following headings.

(i) Loss or damage to the contract works

This covers loss or damage to the construction work itself and any unfixed plant or materials on site yet to be incorporated into the works.

(ii) Public liability

This is to cover the insured against any death, injury or damage claims from members of the public other than their own employees.

(iii) Employer's liability

In many countries, employers of staff have a statutory obligation to provide cover for their employees in the event of death, injury or damage caused during the course of their employment.

(iv) Professional Indemnity insurance

This covers parties' liabilities, particularly in respect of design and the giving of advice.

8.3 Limitation of liability

Clause 80.1 deals with aspects of the Contractor's liability. The first sentence limits liability for loss of or damage to the Employer's property. The limiting amount is to be stated in the Contract Data. Careful thought needs to be given to the limit specified as much will depend on the value, likelihood and consequence of damage, and also proximity of the Employer's property which may be at risk.

It seems that the third insurance in the Insurance Table must include for damage to the Employer's property, therefore insurance cover should be at least for the amount stated as the limit under this clause. Any loss/damage to the Employer's property which is greater than the limit stated will be at the risk of the Employer. A prudent Employer will ensure that such risk is covered by his own insurance of the property.

The second sentence further limits the Contractor's liability to 'direct' losses and does not include such 'consequential' losses as loss of revenue or profit.

The third sentence makes clear that the exclusions and liabilities in this clause apply to liabilities which occur not only in contract but also in tort and other areas of the law.

8.4 Employer's risks within the contract

The following are the Employer's risk within the contract:

1. Risks relating to the Employer's use or occupation of the Site, negligence, breach of statutory duty or interference with any legal right, or a fault in his design. In respect of design, the Employer should cover the risk himself through Professional Indemnity (PI) insurance, or if he uses external consultants for the design, he should ensure that they hold such insurance. This insurance should be held for the full period of liability, which will normally be several years after Completion of the works.
2. Risks relating to items supplied by the Employer to the Contractor until the Contractor has received them, or up to the point of handover to the Contractor. The Employer should ensure that he has adequate insurance in this respect, or again ensure that others who supply items have such insurance.
3. Risks relating to loss or damage to the works, Plant and Materials, caused by matters outside the control of the parties.
4. Risks arising once the Employer has taken over completed work, except a Defect that existed at take-over, an event which was not an Employer's risk or due to the activities of the Contractor on site after take-over.

5.   Risks relating to loss or wear or damage to parts of the works taken over by the Employer, and any Equipment, Plant and Materials retained on site after termination.
6.   Any other risks referred to in the Contract Data.

Any risks not carried by the Employer are carried by the Contractor from the starting date until the issue of the Completion Certificate, the Defects Certificate or a Termination Certificate where appropriate from the Insurance Table.

The Contractor is also obliged, until the Defects Certificate has been issued and unless instructed by the Project Manager, to replace loss of and repair any damage to the Works, Plant or Materials promptly.

8.5 The Insurance Table

The four insurances listed within the Insurance Table are as follows.

(i) Loss of or damage to the works

This first insurance in the Insurance Table covers loss or damage to the project ('the works') as it is being built.

Some other contracts provide an option for either the Contractor or the Employer to provide this insurance, particularly where the works relate to an existing structure or building, where the Employer may already have an existing insurance policy, but the default in the ECSC is that it is provided by the Contractor.

The insurances within the Insurance Table are all effected as a Joint Names Policy, this particular insurance being effective until the Employer's Completion Certificate has been issued and including full reinstatement, replacement or repair (the amount stated should also include an amount or percentage to cover Professional Fees).

The minimum amount of cover for each of the insurances is defined by the Employer within the Contract Data.

Insurance of the works will normally be covered by the Contractors' All Risks (CAR) policy (see section 8.7 below). The minimum cover for all insurances is stated in the Insurance Table; however, the Contractor is liable for whatever the amount of any claim. Therefore he must consider the minimum amount of cover or minimum limit of indemnity in the Insurance Table purely as a guide.

(ii) Loss or damage to Equipment, Plant and Materials

The second insurance covers loss or damage to Equipment, Plant and Materials. Again, this insurance is effected as a Joint Names Policy, but this and the third and fourth insurances below are required to be in force providing cover until the Defects Certificate has been issued.

Again, the Contractor's CAR policy should at least partly cover this. Again, the insurance will normally cover 'replacement cost', meaning the cost of replacement with Equipment, Plant and Materials of similar age and condition rather than 'new for old'. Note that there may be an overlap with this insurance in that Plant and Materials will form part of the works.

(iii) Loss or damage to property (except the works, Plant or Materials) and for bodily injury or death of a person not an employee of the Contractor

This third insurance requires the Contractor to indemnify the Employer against any loss, expense and claim in respect of any personal injury or death caused by the carrying-out of the work, other than his own employees. This includes the liability towards members of the public who may be affected by the construction work, although they have no part in it. In the case where a party makes a claim directly against the Employer due to a death or injury, the Contractor should either take on that claim, or alternatively the Employer can sue the Contractor to recover any monies.

Again, the Insurance Table states the minimum amount of cover or minimum limit of indemnity. The Contractor could be liable for whatever the amount of any claim; therefore he must consider the minimum value in the Insurance Table purely as a guide.

The Contractor also used to have to prove that his Subcontractors had the same cover. However this has now been amended as Contractors can either ensure that the Subcontractors have the cover, or can include it under their own cover on behalf of their Subcontractors.

(iv) Death of or bodily injury to employees of the Contractor

This fourth insurance covers the Contractor's liabilities as an employer of people to insure against injury or death caused to those people whilst carrying out their work, which is a legal obligation in most countries.

8.6 Insurance policies

The Contractor should submit to the Employer for acceptance certificates of insurance as required by the contract:

- before the starting date;
- on each renewal of the insurance policy.

If the Contractor does not submit a required certificate, the Employer may insure a risk, the cost of the premium being paid by the Contractor.

If, as an alternative, the Employer is to provide insurance, then the Employer submits policies and certificates to the Contractor for acceptance before the starting date and as and when instructed by the Contractor.

If the Employer does not submit a required certificate, the Contractor may insure a risk, the cost of the premium being paid by the Employer.

8.7 Contractors' All Risks Insurance

CAR insurance normally covers damage to property, such as damage to buildings and other structures being constructed or to the existing building in which the construction is being carried out. It also normally covers liability for third-party claims for injury and death or damage to third-party property.

The basic principle is that CAR covers those losses not covered by an 'Excluded Risk'. For example, the Joint Contracts Tribunal contracts exclude from cover risks such as Defects due to 'wear and tear, obsolescence, deterioration, rust and mildew, loss and damage arising out of war and for faulty workmanship and faulty design'. The benefit to the insured under this type of policy is that the burden is shifted to the insurer who is required to show that the cause of the loss falls within an exclusion.

This insurance is usually taken out in the joint names of the Contractor and the Employer. Other interested parties, such as funders, often ask to be added as a joint name. The theory is that, if damage occurs to the insured property then, regardless of fault, insurance funds will be available to allow for reinstatement.

The effect of Joint Names Insurance is that each party has its own rights under the policy and can therefore claim against the insurer. Each insured should comply with the duties of disclosure and notification.

8.8 Insurance of neighbouring property

Some contracts have provision for insuring neighbouring property which may suffer from collapse, subsidence or vibration due to damage by, for example, piling, deep excavations or demolition works, where not caused by the negligence of the Contractor.

The ECSC does not provide for this, but it can be added as an additional condition if applicable.

The Employer should consider his liabilities in respect of neighbouring owners who may be affected by the works and, if necessary, either instruct the Contractor to provide such insurance or take it out himself.

Such a policy would normally exclude:

- damage caused by the negligence of the Contractor or his Subcontractors;
- damage caused by errors or omissions in Employer's design;
- damage which can be reasonably foreseen, e.g. cracking may have already occurred to a neighbouring building prior to the works commencing – in this respect, it is important that a condition survey with photographic evidence be carried out before construction commences;
- where the damage is caused by Excepted Risks, these risks will normally be held by the Employer.

8.9 Professional Indemnity or Professional Liability insurance

The ECSC does not provide for PI Insurance to be provided by the Contractor, but certainly if the Contractor is carrying out all or part of the design, it would be advisable for the Employer to include a requirement for PI insurance.

If a Contractor or Consultant providing a service to an Employer makes a mistake, is found to be negligent or gives inaccurate advice, then he will be liable to the Employer in the event that the Employer incurs a loss as a result. This loss can be very significant where the design has to be corrected, parts of the structure have to be taken down and reinstated, a facility has to be closed down whilst the remedial measures take place and there are also legal costs.

PI claims can arise where there is negligence, misrepresentation or inaccurate advice which does not give rise to bodily injury, property damage or personal injury, but does give rise to some financial loss.

Additional coverage for breach of warranty, intellectual property, personal injury, security and cost of contract can be added.

In that event, although the Employer claims against the Contractor or Consultant rather than from the insurers, PI insurance protects the Contractor or Consultant against claims for loss or damage made by a Client or third party.

PI insurance policies are based on a 'claims made' basis, meaning that the policy only covers claims made during the policy period when the policy is 'live', so claims which may relate to events occurring before the coverage was active may not be covered. However, these policies may have a retroactive date which can operate to provide cover for claims made during the policy period but which relate to an incident after the retroactive date. The Employer should ensure that the Contractor has taken out and maintained the required insurances for the full period of his liability.

9 Termination

9.1 Introduction

Termination, including reasons, termination procedures and payment, is covered within the Engineering and Construction Short Contract (ECSC) by Clauses 90.1 to 92.3.

Termination is a word most commonly used in the context of construction contracts to refer to the ending of the Contractor's employment. The parties have a common law right to bring the contract to an end in certain circumstances, but most standard forms give the parties additional and express rights to terminate upon the happening of specified events or matters.

Some contracts refer to termination of the contract, whilst others refer to the termination of the Contractor's employment under the contract. In practice, it makes little difference, and most contracts make express provision for what is to happen after termination.

It should be noted that most contracts require a notice to be given by one party to the other unless the breach is due to insolvency. The offending party then has a period of time in which to remedy the breach, failing which termination can take place.

It is important to note that neither party should attempt to terminate employment under the contract unless in practical terms they have exhausted all other alternatives, they are certain that they wish to do it and also they are sure that the provision is available within the contract. If termination is held to be wrongful, it is usually a repudiation of the contract. Termination is an entitlement, not an obligation.

9.2 Termination within the ECSC

Clause 90.1 covers the procedure to be followed where one Party wishes to terminate. Either Party may notify the other Party, giving details of the reason for termination. The Employer then issues a termination certificate if the reason complies with the contract. The form of the termination certificate is not expressly stated within the contract, so one assumes that the termination is merely recorded in writing. The Contractor does no further work once the termination certificate is issued.

Within the ECSC, there are eight reasons listed for either party to terminate:

- *Reason 1* provides for either party to terminate due to the insolvency of the other. This includes bankruptcy, appointment of receivers, winding-up orders and administration orders dependent on whether the party is an individual, a company or a partnership.
- *Reasons 2 to 5* provide for the Employer to terminate if the Employer has notified the Contractor that he has defaulted and the Contractor has not remedied the default within two weeks of the notification in respect of the Contractor substantially failing to comply with the contract (*reason 2*), substantially hindering the Employer (*reason 3*) or substantially breaking a health or safety regulation (*reason 4*).

  It is quite subjective as to what the words 'substantially failing to comply', 'substantially hindering the Employer' or 'substantially breaking a health or safety regulation' mean. How would one substantially (or unsubstantially) hinder someone, or substantially (or unsubstantially) break a health or safety regulation? Is not a health or safety regulation either 'broken' or 'not broken'?

  The Employer may also terminate for any other reason (*reason 5*).
- *Reasons 6 and 7* provide for the Contractor to terminate if the Employer has not paid an amount due within ten weeks of the assessment day that followed receipt of a Contractor's application, or if the Employer has instructed the Contractor to stop or not start any substantial work or all work for a reason which is not the Contractor's fault and an instruction allowing the work to restart or start has not been given within eight weeks.
- *Reason 8* provides for the Employer to terminate if an event which the Parties could not reasonably have prevented has substantially affected the Contractor's work for a continuous period of more than thirteen weeks. Again, it is quite subjective as to what the words 'substantially affected the Contractor's work' actually mean.

Note that the Contractor can only terminate for one of the above specific reasons, though the Employer can terminate for any reason (reason 5 – Clause 90.3).

Whilst this may seem inequitable, if the Employer terminates for a reason not stated in the contract, he will have to pay the Contractor an additional amount, as stated in Clause 92.3 (see below). This is essentially a loss of profit/overheads provision. Clearly, the earlier the Employer exercises this right of termination, the greater this amount.

9.3 Procedures on termination

Under any of the reasons, whoever terminates, and whether the reason is included in the Termination Table or not, the Employer may complete the works, either himself, or employ others to do so. The Contractor leaves the site and removes his Equipment (Clause 91.1).

9.4 Payment on termination

Also, under any of the reasons, whoever terminates, and whether the reason is included in the Termination Table or not, on termination, the Contractor is paid:

* an amount due assessed as for normal payments;
* the cost of Plant and Materials provided by the Contractor which are on the site or of which the Contractor has to accept delivery;
* any amounts retained by the Employer.

In addition:

* if the Employer terminates for reasons 1, 2, 3 or 4, the amount due on termination includes a deduction of the forecast additional cost to the Employer of completing the works;
* if the Contractor terminates for reasons 1, 6 or 7, or if the Employer terminates for reason 5, under Clause 92.3, the amount due includes 5 per cent of any excess of a forecast of the amount due at Completion had there been no termination over the amount due on termination assessed as for normal payments.

9.5 Termination and Subcontractors

The parties should also consider how termination of their contract affects Subcontractors.

If the Contractor is the defaulter, termination occurs and the Employer wishes to proceed with the Works, any subcontracts would normally be assigned to the Employer.

However, if the Contractor is the terminating party he should ensure that provisions are in place regarding termination of subcontracts, and termination provisions in the main contract should be mirrored in the subcontracts.

10 Disputes

10.1 Introduction

Dispute resolution is provided for within the Engineering and Construction Short Contract (ECSC) by Clauses 93.1 to 93.4, where the UK Housing Grants, Construction and Regeneration Act 1996 are not applicable, and Clause 94.1, where it is applicable.

Note that the Housing Grants, Construction and Regeneration Act 1996 has now been supplemented by the Local Democracy, Economic Development and Construction Act 2009.

If the ECSC is used within the UK, the Employer must by reference to the Act(s) determine whether the contract is a 'construction contract', and select the appropriate dispute resolution option. If the ECSC is used outside the UK, Clauses 93.1 to 93.4 will always apply.

Clearly, there is nothing to prevent the parties jointly attempting to resolve a dispute by a consensual and non-binding route such as negotiation, mediation, conciliation or mini-trial prior to referring it to adjudication.

10.2 What is adjudication?

Adjudication has been defined in the Housing Grants, Construction and Regeneration Act 1996 as

> a summary non-judicial dispute resolution procedure that leads to a decision by an independent person that is, unless otherwise agreed, binding upon the parties for the duration of the contract, but which may subsequently be reviewed by means of arbitration, litigation or by agreement.

It is therefore an intermediate dispute resolution process which is binding upon the parties unless and until the matter is referred to arbitration or litigation.

In the UK, adjudication is a statutory right within a construction contract, but notwithstanding that, it has always been included as the primary means to resolve a dispute within the NEC3 contracts, failing which the parties may refer a dispute to a tribunal. In fact, the majority of adjudication decisions are accepted by parties as the final result and are never taken to the tribunal.

10.3 Who can be an Adjudicator?

Whilst many professional bodies hold registers of Adjudicators there is no formal qualification to become an Adjudicator, though there are commonly acceptable requirements, which include:

- a recognised professional qualification in a relevant discipline. The relevant discipline then normally relates to the dispute in question. For example, if the dispute is related to services installations, mechanical and electrical engineering would probably be a relevant discipline. However it must be remembered that the Adjudicator's role is to enforce the contract, so in that case he does not necessarily have to be a mechanical and electrical engineer just because the dispute is related to services;
- substantial number of years' experience within the construction industry, normally ten to fifteen years post-qualification in a relevant discipline;
- a qualification, or at least a thorough understanding of contracts and legal principles, though in most cases the Adjudicator is not a lawyer, nor does he have to be. Many say it is better if he is not a lawyer, but that is probably a view left to the cynics!
- recognised training in the law and practice of adjudication;
- membership of a professional institution;
- some also say that the Adjudicator should have Professional Indemnity (PI) insurance, though it must be emphasised that, as the Adjudicator is not liable to the parties for any action or inaction on his part other than in bad faith, it could be questioned as to whether such insurance should be a prerequisite to appointment.

As the use of adjudication in the resolution of contractual disputes has grown, particularly in the UK where parties have a statutory right to refer a dispute to adjudication, many education establishments and professional bodies now provide diploma and certificate courses in adjudication.

10.4 Who pays for the adjudication?

The cost of the Adjudicator's fee is shared between the parties. The Adjudicator normally decides what proportion is to be paid by each party, in some cases with one party paying the majority or even all of the Adjudicator's fee. However, both parties are jointly and severally liable for payment of the Adjudicator's fee. Therefore if one party will not or cannot pay his share, possibly due to insolvency, the other party must pay it, and recover the amount directly from the defaulting party.

Each party bears his own costs in collating, copying and postage, which is different to, for example, litigation where both parties' costs are usually borne by the losing party.

10.5 What will the Adjudicator charge?

There are no set hourly rates for Adjudicators. Each Adjudicator will advise the party of his hourly rate and whether there are other costs payable, for example travel or printing. The Adjudicator is required to record properly the time spent carrying out his duties.

10.6 What if a party does not comply with the Adjudicator's decision?

If a party does not comply with the Adjudicator's decision, the other party can enforce that decision by going to court.

10.7 Selecting the Adjudicator

There is provision within the Contract Data for naming the Adjudicator for the contract, including his contact details. Tendering Contractors are then aware at the time of tender who will be the Adjudicator in the event that their tender is successful and a dispute arises and, if necessary, can object to the named person within their tender. Also, the Adjudicator is already in place should a dispute arise. The named Adjudicator may require some form of retainer fee for being named in the contract and being available in the event that a dispute arises.

In practice, many Employers prefer to take one of two other alternative options:

1. The parties can mutually agree to the name of the Adjudicator in the event that a dispute arises. This has the advantage that there is no 'Adjudicator in waiting' and therefore no fee payable. Many practitioners say that pre-appointing the Adjudicator within the contract is resigning oneself to the fact that there will at some point be a dispute. However, once parties are in dispute they then have to agree who is to be the Adjudicator and appoint him; at this point, though, they may not wish to agree with each other about anything!

2. The name of an Adjudicator Nominating Body may be stated by the Employer in the Contract Data. This is probably the favoured option as an independent name can be put forward by the nominating body, who is normally well prepared to put forward the name of an Adjudicator and have him appointed within the time scales set by the contract and if necessary the appropriate legislation.

   In all cases, the Adjudicator must be impartial, i.e. the Adjudicator should not show any bias towards either party. In addition, all correspondence from the Adjudicator must be circulated to both parties.

   Any request for an Adjudicator must be accompanied by a copy of the notice of adjudication, and the appointment of the Adjudicator should take place within seven days of the submission of the notice of adjudication to the other party.

10.8 Adjudicator Nominating Bodies

Adjudicator Nominating Bodies, as the name suggests, are organisations that fulfil the role of nominating Adjudicators. These bodies keep registers of Adjudicators with the required expertise and based at various geographical locations who can act for parties should they be nominated.

A list of recognised Adjudicator Nominating Bodies in the UK is as follows:

- Association of Independent Construction Adjudicators (AICA);
- Centre for Effective Dispute Resolution (CEDR);
- Chartered Institute of Arbitrators (CIArb);
- Chartered Institute of Arbitrators (Scotland) (CIArb-Scotland);
- Chartered Institute of Building (CIOB);
- Construction Conciliation Group (CCG);
- Construction Confederation (CC);
- Construction Industry Council (CIC);
- Construction Plant-Hire Association (CPA);
- Dispute Board Federation (DBF);
- Dispute Resolution Board Foundation (DRBF);
- Institution of Chemical Engineers (IChemE);
- Institution of Civil Engineers (ICE);
- Institution of Electrical Engineers (IEE);
- Institution of Mechanical Engineers (IMechE);
- Law Society of Scotland (LawSoc (Scot));
- Nationwide Academy of Dispute Resolution (NADR);
- RICS Dispute Resolution Service (RICS DRS);
- Royal Incorporation of Architects in Scotland (RIAS);
- Royal Institute of British Architects (RIBA);
- Royal Institution of Chartered Surveyors (RICS);
- Royal Society of Ulster Architects (RSUA);
- Technology and Construction Court Bar Association (TECBAR);
- Technology and Construction Solicitors Association (TeCSA).

Clearly, in naming the Adjudicator Nominating Body in the contract it is advisable to name an organisation with expertise in the project to be carried out.

These bodies are very knowledgeable about appointment of Adjudicators and the relevant time scales and for a modest fee can nominate a suitably qualified Adjudicator to suit the parties' requirements.

10.9 Preparing for adjudication

Once the adjudication process starts by the issuing of a notification of dispute by the referring party, it is a very quick procedure. Therefore it is critical that the referring party, and in turn the responding party, prepare well in the short time they have available. Time scales are stated within the contract, though these may be extended by agreement between the parties.

As the process does not start until the referring party initiates the process by issuing the notification of the dispute to the other party, the referrer does have the advantage of time, and in many cases surprise, as he can prepare his case before issuing the notification.

Before considering initiating the adjudication process, parties should consider the following:

- There must be a dispute. Whilst this may seem fairly obvious, it is essential that a dispute actually exists between the parties in that one party has made a claim which the other party denies (probably repeatedly) and that the parties have exhausted the means to resolve the dispute themselves, for example by the use of negotiation, mediation or conciliation, though they do not have to prove that they have tried other methods.
- The dispute must arise under the contract. The Adjudicator is appointed to decide a dispute under the contract, therefore it is essential that the dispute arises on a matter covered by the contract. If the dispute is whether a contract exists in the first place, then an Adjudicator cannot decide that.
- The role of the Adjudicator is to enforce the contract, not to decide who is the more deserving party. It is therefore essential that, before commencing adjudication, the referring party ensures that he has properly complied with the contract; otherwise he may find himself in a weak position.

   There have been adjudications in the past where a Contractor has acted in good faith but not strictly in accordance with the contract, accepting and willingly acting on verbal instructions in order not to damage what appeared to be a good relationship with the Employer. Then, when a dispute arises the Adjudicator states that as the contract did not provide for verbal instructions, the Contractor was incorrect to act on them, and therefore the Contractor loses the adjudication.
- Whilst many people say that, as adjudication is a fairly simple process, it should not include lawyers as the issues are normally construction-related rather than legal issues, it is important that the parties carry out their obligations within the short time scale they have. So if a party is uncertain as to what they have to do, and when and how they have to do it, they obtain professional advice, which may be from a lawyer.
- It is essential that a clear and concise notice of adjudication is prepared by the referring party as it defines what is in dispute, and the matters that are being referred to the Adjudicator. The Adjudicator cannot consider issues which are not set out within the notice of adjudication.

The notice should include:

- brief details of the project. Whilst the Adjudicator may have been named in the contract, he may have little knowledge about the actual project;
- the names and addresses of the parties to the contract and the parties in dispute;

- the nature and a brief description of the dispute;
- the nature of the redress which is sought from the Adjudicator. This is critical: many referring parties do not make it clear what they want the Adjudicator to decide!

Details included within the referral should be clearly and concisely presented so that the responding party knows what it needs to respond to, and also the Adjudicator clearly understands what is required of him. Any supporting information should be clearly referenced and relative to the matter in dispute. Neither party should include or rely on any information or evidence which the other party has not seen. If they do, the other party may raise an objection, which could bring the adjudication process to a premature end without a resolution.

The contract provides for time scales to be extended by agreement between the parties, so if more time is needed in order to prepare and submit documentation, the other party should be consulted.

10.10 Statutory right to adjudication (UK-based contracts)

When the Housing Grants, Construction and Regeneration Act 1996 was introduced in the UK, it applied to construction contracts which came into existence after 1 May 1998, and, apart from a few notable exceptions, it provided a statutory right to refer a dispute under a construction contract to adjudication. This Act has now been supplemented by the Local Democracy, Economic Development and Construction Act 2009.

10.11 What do the Housing Grants, Construction and Regeneration Act 1996 and the Local Democracy, Economic Development and Construction Act 2009 cover?

The Act applies to all construction contracts, including:

- all normal building and civil engineering work, including construction, alteration, repair, maintenance, extension and demolition or dismantling of structures forming part of the land and works forming part of the land, whether they are permanent or not;
- the installation of mechanical, electrical and heating works and maintenance of such works;
- cleaning carried out in the course of construction, alteration, repair, extension, painting and decorating and preparatory works;
- agreements with consultants such as Architects, other Designers, Engineers and Surveyors;
- elements such as scaffolding, Site clearance, painting and decorating;
- labour-only contracts;
- contracts of any value.

It excludes:

- work on process plant and on its supporting or access steelwork on sites where the primary activity is nuclear processing, power generation, water or effluent treatment, handling of chemicals, pharmaceuticals, oil, gas, steel or food and drink;
- supply-only contracts, that is, contracts for the manufacture or delivery to site of goods where the contract does not provide for their installation;
- extracting natural gas, oil and minerals (and the workings for them);
- purely artistic work;
- off-site manufacture;
- contracts with residential occupiers – this means a contract where work is carried out on a dwelling which one of the parties occupies or intends to occupy as his residence rather than simply a housing project;
- private finance initiative (PFI) contracts – this only includes the head agreements in PFI projects, not a contract for the construction of the Works;
- finance agreements – this includes loan agreements.

The Adjudicator is appointed jointly by both parties who are jointly and severally liable for those fees.

Under the Act, the contract must include an adjudication clause which complies with the requirements of the Act. If the contractual clause does not comply then it is voidable and the parties can exercise their right to adjudication under the Scheme for Construction Contracts issued by the Secretary of State.

The provisions in the contract must:

- enable the parties to give notice at any time;
- provide a timetable with the object of securing the appointment of the Adjudicator and referral of the dispute to him within seven working days;
- require the Adjudicator to reach a decision within 28 days of referral or such longer period as is agreed by the parties after the dispute has been referred;
- allow the Adjudicator to extend the period of 28 days by up to 14 days with the consent of the party by whom the dispute was referred;
- impose a duty on the Adjudicator to be impartial;
- enable the Adjudicator to take the initiative in ascertaining the facts and the law;
- provide that the decision of the Adjudicator is binding until determined by litigation, arbitration or by agreement;
- provide that the Adjudicator is not liable for anything done or omitted in the discharge or purported discharge of his functions as Adjudicator unless the act or omission is in bad faith, and that any Employee or agent of the Adjudicator is similarly protected from liability.

N.B.: In September 2011, the NEC3 publishers issued a brief amendment to the ECSC adjudication provisions to align it with the Local Democracy, Economic Development and Construction Act 2009.

The following replaces subclause 93.3(1):

- A Party may issue to the other Party a notice of his intention to refer a dispute to adjudication at any time. He refers the dispute to the *Adjudicator* within one week of the notice.
- The Adjudicator may in his decision allocate his fees and expenses between the Parties.
- The Adjudicator may, within five days of giving his decision to the Parties, correct the decision to remove a clerical or typographical error arising by accident or omission.
- If the Adjudicator's decision changes an amount notified as due, payment of the sum decided by the Adjudicator is due not later than seven days from the date of the decision or the final date for payment of the notified amount, whichever is the later.

This amendment was then embodied into the April 2013 reprint.

10.12 Does a 'dispute' exist?

Only a dispute between the contracting parties can be referred to adjudication. If there is no dispute, then an Adjudicator has no jurisdiction to consider the matter.

As an example, on a non-NEC contract, if the Contractor submits a loss and expense claim and the Contract Administrator does not agree with it, then the simple fact that the claim is not agreed does not mean that there is a dispute as the parties are still free to meet and, if necessary, to negotiate a settlement. It is only when the parties have exhausted all the opportunities to meet and reach a compromise, but still have a difference, that a dispute exists, and the matter can be referred to the Adjudicator.

It must also be noted that an Adjudicator only has jurisdiction to consider one dispute at a time under the same contract unless there is agreement between the disputing parties. Therefore, if there are a number of issues in dispute, unless they are inextricably joined then there are likely to be several adjudications.

Disputes under or in connection with the contract may be referred to the Adjudicator. Note that a dispute under the contract cannot be referred to the tribunal (arbitration or litigation) unless it has first been decided by the Adjudicator.

Adjudication is a quick, convenient and relatively inexpensive way of resolving a dispute, whereby an impartial third-party Adjudicator named in the contract, selected by agreement between the parties or nominated by a third party decides the issues in dispute between the parties. It is quicker and less expensive than arbitration or litigation, though the parties may refer to one of those methods following adjudication.

It must be recognised that the role of the Adjudicator is to enforce the contract, not to decide what, in his opinion, would be the fairest outcome. For example, if the parties have a dispute over the quality of suspended ceilings carried out, it is the Adjudicator's task to ascertain whether the ceilings were supplied and installed in accordance with the contract, i.e. drawings and specifications, and the applicable law, not whether it was a good or bad job. In addition, the question of whether a contract exists cannot be decided by the Adjudicator but more likely through the courts.

10.13 Adjudication provisions within the ECSC

Clauses 93.1 to 93.4 describe how disputes between the Parties are resolved initially by adjudication.

The ECSC provides two options for referral to the Adjudicator:

(i) If the Housing Grants, Construction and Regeneration Act 1996 amended by the Local Democracy, Economic Development and Construction Act 2009 does not apply to this contract

The Adjudicator

Clause 93.2 requires the Parties to appoint the Adjudicator under the NEC Adjudicator's Contract current at the starting date.

The Adjudicator is appointed under the NEC3 Adjudicator's Contract, he acts impartially, and if he resigns or is unable to act the parties jointly appoint a new Adjudicator. One would assume that the referring party obtains a copy and completes the Adjudicator's Contract. If the parties have not appointed an Adjudicator, either party may ask the nominating body to choose an Adjudicator within four days of the request and also what the parties do if the Adjudicator resigns or is unable to act.

No procedures have been specified for appointing a suitable person, and in practice a number of different methods have been used. Whatever method is used, it is important that both parties have full confidence in his impartiality, and for that reason it is preferable that a joint appointment is made. The Adjudicator should be a person with experience in the type of work included in the contract between the parties and who occupies or has occupied a senior position dealing with disputes. He should be able to understand the viewpoint of both parties.

Often the parties delay selecting an Adjudicator until a dispute has arisen, although this frequently results in a disagreement over who should be the Adjudicator.

As noted, the selection of the Adjudicator is important, and it should be recognised that a failure to agree an Adjudicator means that a third party will make the selection without necessarily consulting the parties.

The referring party must include within his referral information which he wishes to be considered by the Adjudicator. Any more information from either party is to be provided within four weeks of the referral.

A dispute under a subcontract, which is also a dispute under the ECSC, may be referred to the Adjudicator at the same time and the Adjudicator can decide the two disputes together.

The Adjudicator may review and revise any action or inaction of the Parties, take the initiative in ascertaining the facts and the law relating to the dispute, instruct a Party to provide further information and instruct a Party to take any other action which he considers necessary to reach his decision.

All communications between a party and the Adjudicator must be communicated to the other party at the same time.

The adjudication

Either Party must notify a dispute to the other Party within four weeks of becoming aware of the dispute, and then refer it to the Adjudicator between two and four weeks after that notification. These provisions are time-barred, so neither Party may refer a dispute to the Adjudicator unless they have complied with the time scales.

The Party referring the dispute is required to include within his referral information to be considered by the Adjudicator. Any additional information is provided to the Adjudicator within a further two weeks of the initial referral, though this period can be extended by agreement with the Adjudicator and the other Party.

Any communication between a Party and the Adjudicator must at the same time be copied to the other Party.

The Adjudicator decides the dispute and notifies the parties and his reasons within four weeks of the referral. This four-week period may be extended with the consent of the referring Party or by any period agreed by the Parties.

If the Adjudicator does not give his decision within the required time, the Parties act as if the Adjudicator has resigned.

Until this decision has been communicated, the parties proceed as if the matter in dispute was not disputed.

The Adjudicator's decision is binding on the parties unless and until revised by a tribunal and is enforceable as a contractual obligation on the parties. The Adjudicator's decision is final and binding if neither party has notified the Adjudicator that they are dissatisfied with an Adjudicator's decision within the time stated in the contract, and intends to refer the matter to the tribunal.

Review by the tribunal

A dispute cannot be referred to the tribunal unless it has first been referred to the Adjudicator.

The tribunal is named by the Employer within the Contract Data. Whilst no alternatives are stated, it will normally be litigation or arbitration. If arbitration is chosen the Employer must also state in the Contract Data the arbitration procedure.

A party can, following the adjudication, notify the other party within four weeks of the Adjudicator's decision that he is dissatisfied. This is a time-barred right, as the dissatisfied party cannot refer the dispute to the tribunal unless it is notified within four weeks of the Adjudicator's decision. Failure to do so will make the Adjudicator's decision final and binding.

The tribunal settles the dispute and has the power to reconsider any decision of the Adjudicator and review and revise any action or inaction of the Employer or Contractor.

It is important to note that the tribunal is not a direct appeal against the Adjudicator's decision, the parties have the opportunity to present further information or evidence that was not originally presented to the Adjudicator, and also the Adjudicator cannot be called as a witness.

(ii) If the Housing Grants, Construction and Regeneration Act 1996 applies to this contract

The inclusion of adjudication within the New Engineering Contract, now the NEC3, pre-dates UK legislation giving parties to a contract the statutory right to refer a dispute to adjudication.

The Local Democracy, Economic Development and Construction Act 2009

The Local Democracy, Economic Development and Construction Act 2009, which came into force in England and Wales on 1 October 2011 and in Scotland on 1 November 2011, amended the Housing Grants, Construction and Regeneration Act 1996, the new Act applying to most UK contracts after that date. There still remain certain categories of contract to which neither Act applies.

The differences between the Housing Grants, Construction and Regeneration Act 1996 and the Local Democracy, Economic Development and Construction Act 2009 have been explored in the Introduction, but in respect of adjudication they include:

* The earlier Act only applied to contracts which were in writing, whereas the new Act also applies to oral contracts.
* Terms in contracts such as 'the fees and expenses of the Adjudicator as well as the reasonable expenses of the other party shall be the responsibility of the party making the reference to the Adjudicator' have been prohibited. Much has been written about the effectiveness of such clauses, often referred to as Tolent clauses (*Bridgeway Construction Ltd v Tolent Construction Ltd 2000*), and whether they comply with the earlier Act, so the new Act should provide clarity for the future.
* The Adjudicator is permitted to correct his decision so as to remove a clerical or typographical error arising by accident or omission. Previously he could not make this correction.

Section 108 of the new Act provides parties to construction contracts with a right to refer disputes arising under the contract to adjudication. It sets out certain minimum procedural requirements which enable either party to a dispute to refer the matter to an independent party who is then required to make a decision within 28 days of the matter being referred.

If a construction contract does not comply with the requirements of the Act, or if the contract does not include an adjudication procedure, a statutory default scheme, called the Scheme for Construction Contracts (referred to as the 'Scheme'), will apply.

The Act provides that a dispute can be referred to adjudication 'at any time' provided the parties have a contract. This is not necessarily during the construction stage; for example a Designer may refer a dispute during the design stage.

'At any time' can also refer to a dispute after the contract is completed.

The June 2005 version of the contract included Clause 94.1:

> A Party may issue to the other Party a notice of his intention to refer a dispute to adjudication at any time. He refers the dispute to the Adjudicator within one week of the notice.

What are the requirements?

Section 108 of the Act requires all 'construction contracts', as defined by the Act, to include minimum procedural requirements which enable the parties to a contract to give notice at any time of an intention to refer a dispute to an Adjudicator.

The contract must provide a timetable so that the Adjudicator can be appointed, and the dispute referred, within seven days of the notice. The Adjudicator is required to reach a decision within 28 days of the referral, plus any agreed extension, and must act impartially. In reaching a decision an Adjudicator has wide powers to take the initiative to ascertain the facts and law related to the dispute.

The Housing Grants, Construction and Regeneration Act 1996 defines adjudication as:

> a summary non-judicial dispute resolution procedure that leads to a decision by an independent person that is, unless otherwise agreed, binding upon the parties for the duration of the contract, but which may subsequently be reviewed by means of arbitration, litigation or by agreement.

In that sense, adjudication does not necessarily achieve final settlement of a dispute because either of the parties has the right to have the same dispute heard afresh in court, or, where the contract specifies, arbitration.

However, experience since the Housing Grants, Construction and Regeneration Act 1996 came into force shows that the majority of adjudication decisions are accepted by the parties as the final result.

The Adjudicator is appointed under the NEC Adjudicator's Contract, he acts impartially, and if he resigns or is unable to act, the parties jointly appoint a new Adjudicator. If the parties have not appointed an Adjudicator, either party may ask the nominating body to choose an Adjudicator within four days of the request.

A party may first give a notice of adjudication to the other party with a brief description of the dispute, details of where and when the dispute has arisen and the nature of the redress which is sought.

The Adjudicator may be named in the contract, in which case the party sends a copy of the notice to the Adjudicator. The Adjudicator must confirm within three days of receipt of the notice that he is able to decide the dispute, or if he is unable to decide the dispute.

Within seven days of the issue of the notice of adjudication, the party:

- refers the dispute to the Adjudicator;
- provides the Adjudicator with the information on which he relies together with supporting information;
- provides a copy of the information and supporting documents to the other party.

A dispute under a subcontract, which is also a dispute under the ECSC, may be referred to the Adjudicator at the same time and the Adjudicator can decide the two disputes together.

10.14 NEC3 Adjudicator's Contract

Introduction

The first edition of the NEC Adjudicator's Contract was published in 1994, and was written for the appointment of an Adjudicator for any contract under the NEC family of standard contracts. The second edition, published in 1998, contained some changes, including the need to harmonise with the NEC standard contracts and further editions which had been issued since 1994. The third edition harmonises the contract with the NEC3 family of contracts.

The NEC3 Adjudicator's Contract can also be used with contracts other than NEC3, though, depending on the contract and the applicable law, some amendment may be necessary.

In the UK, the Housing Grants, Construction and Regeneration Act 1996 has made adjudication mandatory as a means of resolving disputes in construction contracts which fall under the Act. Parties to a contract which does not provide for adjudication as required by the Act have a right to adjudication under the Scheme for Construction Contracts. Schemes which are substantially similar have been published for England and Wales, Scotland and Northern Ireland.

The contract consists of five pages plus the index. The agreement is between the two disputing parties and the Adjudicator. The contract contains the following four sections.

(i) General

This covers the obligation upon the Adjudicator to act impartially, and to notify the parties as soon as he becomes aware of any matter which could present a conflict of interest.

There is a definition of expenses, which include printing costs, postage, travel, accommodation and the cost of any assistance with the adjudication. All communications must be in a form which can be read, copied and recorded. This is a requirement of all NEC contracts, but also of the adjudication process itself, with both parties and the Adjudicator having to be copied into any communication between the parties.

(ii) Adjudication

The Adjudicator cannot decide any dispute that is the same or substantially the same as his predecessor decided. He must make his decision and notify the parties in accordance with the contract, and in reaching his decisions he can obtain assistance from others, but must advise the parties before doing so, and also provide the parties with a copy of anything produced by the assisting party, so that the parties can be invited to comment. The Adjudicator's decision is to remain confidential between the parties, and following his decision the Adjudicator retains documents provided to him for the period of retention, which is stated in the Contract Data.

(iii) Payment

The Adjudicator's hourly fee is stated in the Contract Data. There is provision for an advance payment to be made by the referring party to the Adjudicator if stated in the Contract Data.

Following his decision, the Adjudicator submits an invoice to each party for their share of the amount due, including expenses. The parties pay the amount due within three weeks of receiving the Adjudicator's invoice. Interest is paid at the rate stated in the Contract Data on any late payment. The parties are jointly and severally liable to pay the amount due to the Adjudicator. Therefore if one party fails to pay, the other must pay, including any interest on the overdue amount, and recover the amount from the other party.

(iv) Termination

The parties may, by agreement, terminate the Adjudicator's agreement. Also, the Adjudicator may terminate the agreement if there is a conflict of interest, he is unable to decide a dispute, an advance payment has not been made or he has not been paid an amount due within five weeks of the date by which the payment should have been made. The Adjudicator's appointment terminates on the date stated in the Contract Data.

10.15 The tribunal

Whilst the Contract Data requires the Employer to enter the form of the tribunal, it does not say what choices are available. Other contracts refer disputes to expert determination or Dispute Review Boards, but under the ECSC the Employer normally selects arbitration or litigation. Within the Contract Data the Employer is required to state the arbitration procedure, the place where the arbitration is to be held and who will choose an Arbitrator.

There are a number of differences between the two, but essentially arbitration is hearing the dispute in private and litigation is hearing the dispute in a public court.

It must be stressed that the role and authority of the tribunal are not to review and appeal against the previous decision of an Adjudicator, but to resolve the dispute independently of the adjudication. Therefore it is quite in order for parties to introduce new evidence or materials to the tribunal.

10.16 Principles of arbitration

Any dispute between two or more parties can be resolved through the courts, a process known as litigation or legal proceeding.

However, litigation has many disadvantages, not least the cost of a full court hearing, but also in many cases a number of years of waiting time before the matter actually reaches the courts, so in recent times arbitration has become increasingly used as a simpler, more convenient method of dispute resolution.

Arbitration is the hearing of a dispute by a third party, who is often not a lawyer, though that term has a very wide meaning, but is an expert in the field in which the dispute is based, and can give a decision based on opinion, which is then legally binding on both parties.

Arbitration is not a new concept. In fact, it has been in existence for almost as long as the law itself. The first official recognition in the UK was the Arbitration Act 1697, which largely governed disputes about the sale of livestock in cattle markets. With the continuing growth of freight transport by ship, and the advent of railways and other forms of transport, a large number of arbitration cases ensued, and as a result Parliament passed the Arbitration Act 1889.

Various amendments were introduced over the following years, with the current legislation being the Arbitration Act 1950, as amended by the Arbitration Acts of 1975 and 1979. The Arbitration Act 1996 is the current legislation related to arbitration in England and Wales.

One does not issue writs in arbitration as one would with litigation: *both* parties agree to enter into arbitration as a pre-agreed provision within a contract and thus be bound by the decision of the Arbitrator.

Arbitrators are appointed by one of three methods:

1. as a result of a pre-agreed decision between the parties, normally through a provision within a contract to go to arbitration in the event of a dispute. In

this case the Arbitrator himself may be named or subject to nomination by a professional body such as the Chartered Institute of Arbitrators;
2. by statute. Legislation in many countries includes the provision for disputes to be settled by arbitration;
3. by order of a court.

There are three arbitration bodies in the UK:

1. the Chartered Institute of Arbitrators;
2. the London Court of International Arbitration;
3. the London Maritime Arbitrators Association.

The Arbitration Act is fairly brief and sets out procedures in the absence of any agreement to the contrary between the parties.

As arbitration is a more flexible arrangement than litigation. Whatever procedure the Arbitrator and both parties agree to is sufficient in an arbitration case.

Whilst most forms of contract provide for resolution of disputes by arbitration, it must be stressed that it should only be seen as a last resort when all attempts at negotiation and other resolution methods have failed.

It may be surprising to hear that arbitration is common as a settlement procedure for disputes under trade agreements, maritime and insurance, consumer matters such as package holidays, construction industry disputes and property valuations.

The contracting party that requested that the dispute be referred to arbitration is referred to as the 'Claimant', whilst the other party is referred to as the 'Respondent'.

Should there be a Joint Agreement to go to arbitration, the Arbitrator will decide who is the Claimant, and who is the Respondent.

It is important that these titles are clarified as the Arbitration Rules repeatedly refer to them.

Advantages of using arbitration

There are several advantages in using arbitration rather than litigation.

(i) Privacy

This is a distinct advantage, as in litigation, private and often personal matters are debated in full view of the public in open court, and often published in the press and technical magazines, whereas arbitration is carried out in private sessions, usually at the Arbitrator's office or at a pre-arranged venue.

(ii) Convenience

Arbitration is carried out at a time and place to suit the parties, for example evenings and weekends, not when the court is available for session.

(iii) Speed

The dispute is settled efficiently without the normal delays and procedural matters involved in often fairly lengthy and formal litigation.

(iv) Simplicity

There are few technical procedures as in the courts, the parties set their own procedures within a set of ground rules.

(v) Expertise

The Arbitrator will be an expert on the matter in dispute, for example a Structural Engineer where the dispute is about structural design issues, whereas judges and other legal professionals are experts only on principles of law. The dispute will often not hinge on a legal problem; for example, in matters of design who better than a Designer to resolve it?

(vi) Thoroughness

Because the Arbitrator is a skilled technical person, the dispute can often be dealt with more thoroughly and matters more rigorously aired than in a confined and formal court atmosphere.

(vii) Expense

It can be readily seen that arbitration is generally a less expensive option than litigation, though it must not be seen as a cheap way to resolve a dispute. In some complex cases arbitration can be even more expensive than litigation.

It is worth mentioning at this point that, in many countries, legal aid, which can be available for litigation, is not available for arbitration.

Disadvantages of using arbitration

The main disadvantages of using arbitration rather than litigation are as follows.

(i) Lack of legal knowledge

A dispute may hinge on a difficult point of law, which the Arbitrator has insufficient legal knowledge to resolve, as he is an expert in the field relating to the dispute, not in the law itself. He is, however, entitled to seek legal advice on a point of law, but not on the dispute itself, and also the disputing parties can refer a question of law to the High Court.

N.B.: The arbitration award itself cannot be referred to the courts unless *both* parties consent, or the courts are referred to on a point of law, the resolution of which could substantially overturn an Arbitrator's decision.

Further, Arbitrators cannot be sued for negligence as they do not hold themselves up as experts in the sense of being advisory bodies and therefore liable for giving wrongful advice.

(ii) Precedents

Arbitration is not subject to rulings in previous cases as with the courts. Each case is judged on its own merits on the day, and also to some extent subject to the personality of the Arbitrator; therefore there is no guide as to how successful an action could be.

It is prudent always to seek professional advice before resorting to arbitration as, whilst it is simpler than court action the Arbitrator's decision is final and binding, therefore a party must be confident that they have a strong case.

(iii) Decision making

It has been said that the Arbitrator, being a technical person and possibly having been involved in a similar dispute himself, tends to be fairer than judges in litigation. Arbitrators also tend to sympathise with both parties and can split the difference between the parties when making their awards as they have possibly been in a similar position themselves, whereas judges are inclined to take colder decisions and award wholly in favour of one of the parties. It has to be said that this is not a commonly held view!

10.17 Reference to arbitration

Although arbitration is a different process to adjudication, the same initial principles apply before arbitration can commence in that there must be a dispute, there must be an agreement to arbitrate, which will normally be provided for within the contract, and a party must submit a notice to initiate the process.

Proceedings begin by either party sending a written notice to the other stating that they wish to refer the dispute to arbitration. The form and process may be dictated by the arbitration procedure stated by the Employer within the Contract Data.

The parties may agree on a person to act as Arbitrator, or a professional body will select and appoint the Arbitrator. In some cases, where the parties cannot agree a choice and there is no provision for a third party to do it, a court may select an Arbitrator. There is normally only one Arbitrator, but in some cases there are three, consisting of two Arbitrators and a third who acts as an umpire. Any correspondence flowing between the two parties as from the date of appointment must then also be copied to the Arbitrator(s).

The next step is that the Arbitrator writes to the parties accepting the appointment. A Preliminary Meeting is held between the parties at a time and venue to be directed by the Arbitrator to decide the form of the hearing, which may be on a documents-only basis or an oral hearing. Documents-only tends to be cleaner and quicker as there is less opportunity for the parties to debate the issue.

10.18 The role of the Arbitrator

The role of the Arbitrator is only to examine the evidence before him to enable him to resolve the dispute, and nothing more. He cannot look into other matters not in dispute. He can conduct the arbitration as he wishes, though he must gear the procedures in a way that is not detrimental to either party.

He must also have a regard for the rules of evidence. For example:

- Hearsay evidence is not permissible.
- He is not permitted to use evidence from someone not qualified to give such evidence.
- The Arbitrator is not allowed to take into account any evidence he has discovered himself, which was not provided by either party.
- He is not allowed to use any privileged evidence, for example that which is marked 'Without Prejudice'.
- In addition, the rules of discovery apply where either party is entitled to have access to the file or documents of the other provided that it relates to matters at issue.

10.19 Inspection by the Arbitrator

The Arbitrator is entitled to inspect any property, work or Materials at any premises, and may invite the Claimant and/or the Respondent to accompany him purely for the purpose of identifying the work or Materials.

No one else is allowed to accompany the Arbitrator unless he specifically invites him.

10.20 The award

The award is binding upon both parties and must be served in the appropriate manner with full headings, summary of issues in dispute, outline of events leading to the dispute, his decision based on the hearing and finally the monetary award.

There is very limited scope for appeal against an Arbitrator's award. Usually, appeals can only be based on a claim that the Arbitrator was wrong on a point of law, or there was serious and proven irregularity.

Arbitration awards can also be enforced in court if necessary.

10.21 Arbitrator's fees and expenses

Both parties will normally incur costs, though it is the Arbitrator's decision as to how the costs are apportioned. Both parties are initially liable to the Arbitrator for payment of his fees and expenses, but the Arbitrator will direct the parties as to the proportion they are liable for once the decision is reached.

Should the parties agree on a settlement of the dispute at any time during the arbitration, they will jointly be liable for the Arbitrator's costs.

10.22 Litigation

It is not proposed to go into any detail about the litigation process from initiation to judgement by a court, first as the contract does not give any details of the process, but also because the proceedings themselves will depend on the law of the contract, which in turn is usually dependent on the location of the project.

However, the process will normally involve the Claimant's representative, usually a Solicitor, issuing a claim form or writ to the appropriate court; papers are then served on the Defendant. The Defendant then replies to the service by either admitting or denying the claim against him. If he denies it, he replies by sending his defence, which may also be accompanied by any counter-claim. The case is then allocated to a court. There then follows a period of accumulation, collation and presentation of evidence, and hearings. Again, the process, the submission and the form of the hearing are dependent upon the law of the contract, until the court passes judgement.

11 Tendering

11.1 Deciding the procurement strategy

Deciding the procurement strategy is the foundation stone to the successful outcome to any contract.

In terms of NEC contracts, this is regardless of whether it is a larger, higher-risk project using the Engineering and Construction Contract (ECC) or a smaller, lower-risk project using the Engineering and Construction Short Contract (ECSC), but this process is very often understated or almost overlooked as Employers and their advisers, in their urgency to get projects out to tender and commenced on site, tend to select the procurement strategy most familiar to them, rather than spending time properly considering the project in terms of the Employer's resources and his objectives in terms of *time*, *price* and *quality*, with the additional factor of *risk* also having to be considered and evaluated. It is rarely achievable for all three to be fully satisfied; it is usually a compromise (Figure 11.1).

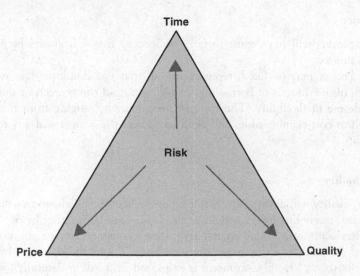

Figure 11.1 The time, price, quality and risk triangle.

The initial briefing with the Employer is the fundamental first step in establishing what is required, when it is required and what the Employer is proposing and/or can afford to pay. Through a process of interviewing key members of the Employer's team and reviewing his business plans, specific objectives, options and constraints, a procurement strategy is developed which forms the basis for future assessments and recommendations.

Whilst it may sound obvious, it is vital that the Employer understands what he wants and can also communicate it clearly. He may not know what he wants, but also he can often be swayed by consultants or other third parties into something he does not want.

In broad terms, the procurement strategy should consider time, price and quality and the balance between the three as follows.

(i) Time

Most Employers want their projects to be completed as quickly as possible. This is particularly relevant in sectors such as retail, where the Employer needs to open a shop as early as possible in order to sell goods. Also, an Employer may need to tie in completion of a project with transfer of staff, students or patients from older buildings. And some projects have absolute deadlines, e.g. major public events and national and international sporting events.

If the Employer sees time as being the most important factor, he may have to pay more money, and quality may have to be lowered in order to achieve that.

The procurement method will have to consider timing through programme requirements, but also through provisions such as sectional completion and delay damages.

(ii) Price

All Employers will have some form of budget, so this will always be a major consideration.

The budget may be fixed, representing all that the Employer has available through his own funds or borrowings in order to fund the project, or may have some degree of flexibility. The form of procurement, ranging from lump sum through to cost-reimbursable, will need to address the budget and any required flexibility.

(iii) Quality

Finally, quality will always be a major factor, though if the above two are more important, many Employers will expect to lower standards either by not having what they want, or by value engineering, which considers similar but more economical solutions. It is important that the design lowers the standards by agreement, not that a 'cheap' Contractor is employed who will unilaterally lower the standards himself!

More detailed questions can then be asked and answered to complete the procurement strategy, which include the following:

- How important to the Employer is certainty of price, particularly if changes are expected? Lump sum options will give greater certainty of price, whereas cost-reimbursable options will give less certainty of price, but greater flexibility.
- What resources and expertise does the Employer have? Cost-reimbursable and target contracts require resources to check invoices and time sheets, as opposed to lump sum or remeasurable contracts.
- Which party is to be responsible for design and/or which party has the necessary design expertise? Design can be assigned to the Contractor, though it is not recommended that bills of quantities are used for design and build.
- How important to the Employer is an early start and/or early completion? Early commencement and early completion may lead the Employer towards a cost-reimbursable solution as lump sum and remeasurement procurement options require a considerable time to prepare the tender documentation for pricing.
- How clearly defined is the Scope of Works, and what form does it take? For example, is it in the form of detailed drawings and specification, or performance-based criteria? Again, the contracts can be used with detailed drawings or performance-based criteria, where the Contractor carries out the design.
- What is the likelihood of change to those defined requirements? Many Employers prefer a remeasurement option where there is a likelihood of many changes. However, the ECSC has a dynamic change management system (compensation events) which allows changes to be priced speedily.
- What is the Employer's attitude towards risk? Who is best placed to manage the inherent risks within the project? Lump sum contracts assign greatest risk to the Contractor; cost-reimbursable contracts assign greater risk to the Employer.
- It is a poorly acknowledged fact that the Employer always pays for risk, regardless of whether it is carried within the contract by the Employer or Contractor. If it is the Contractor's risk he is deemed to have priced and programmed for it; if it is an Employer's risk he will pay for it if it occurs.

The procurement strategy then leads to a contract strategy which considers what is available in terms of contract to enable the Employer to achieve the objectives drawn from the procurement strategy.

Many traditional contracts are based on bills of quantities with the Employer designing the project, and detailed bills of quantities being sent to each tendering Contractor for pricing. However, in the past 20 years there has been a movement away from the use of traditional bills of quantities towards Contractor-designed projects with payment mechanisms such as milestone payments and activity schedules, with payment based on progress achieved, rather than quantity of work done.

11.2 Contractor design

The increasing growth of Contractor design has given rise to the increasing use of various design and build contracts, performance-specified works, and design, build, finance and operate projects.

There are a number of reasons for allocating some or most of the design to the Contractor:

- The design and construction periods can overlap, leading to faster delivery of the project.
- The Contractor can utilise his experience and preferred methods of construction to build rationally designed projects which minimise costs and programme durations.
- The temporary works/permanent works interface and influence of design is rationalised as both remain with the Contractor and therefore the permanent works should be more 'buildable'.
- The traditional design/construction interface and the risks associated with it are transferred to the Contractor.
- The management of the design risk by the Contractor can result in greater certainty of the time, cost and performance and project objectives being met.

There has also been an increasing use in the past 15 years of target contracts which have been encouraged by the increasing use of partnering and collaborative working arrangements, and the associated more equitable sharing of risk.

11.3 Partnering arrangements

The past 20 years, particularly since the publication of the Latham[1] and Egan[2] reports, have seen the growth of partnering and framework agreements.

The US Construction Industry Institute has defined partnering as:

> A long-term commitment between two or more organizations for the purpose of achieving specific business objectives by maximizing the effectiveness of each participant's resources . . . the relationship is based upon trust, dedication to common goals and an understanding of each other's individual expectations and values.[3]

Partnering is a medium- to long-term relationship between contracting parties, whereby the Contractor, and in turn Consultants and various other parties, are not required to tender competitively on price for each project, but are awarded the work on the basis of their ability to deliver, and normally by price negotiation.

The construction industry is known to be a high-risk business, and many projects can suffer unexpected cost and time overruns, frequently resulting in disputes between the parties.

The risks within a project are initially owned by the Employer, who may choose to adopt a 'risk transfer' approach where the risks are assigned through the contract to the Contractor who has the opportunity to price and programme for them, or a 'risk embrace' approach where the Employer retains the risks. In reality, most contracts are a combination of the two.

The traditional approach to risk management is that of risk transfer, which is fine if the Scope of Works is clear and well defined. However, in recent years Employers have become more aware that they can achieve their objectives better by adopting a more 'old-fashioned' risk embrace culture.

Advantages of partnering agreements

- The procurement process from conception to commencement on site, and subsequent completion, can be significantly reduced in terms of time and resources.
- There is no re-tendering time and cost for future projects.
- Relationships can be developed based on trust.
- Contractors are appointed earlier and can contribute to the design and procurement process through Early Contractor Involvement (ECI).
- There tends to be greater cost certainty.
- Continuous improvement can be achieved by transferring learning from one project to the next.
- Better working relationships can be developed as the parties know each other.
- Prices can be more competitive and resources used more efficiently by continuous flows of work.

Disadvantages of partnering agreements

- Obligations and liabilities can become less clear in time as the parties can lapse into informal working patterns.
- Complacency can set in after some time as the Contractor has an assured flow of work.
- Employers are often dissatisfied that they are not getting value for money in their projects.
- Continuing to award the work to a small number of Contractors, or even one Contractor, prevents other equally good, or better, Contractors having the opportunity to carry out work.

All of these disadvantages can be overcome by maintaining a disciplined approach to communications between the parties and their rights and obligations, and also by continuous measurement of performance and deliverables.

11.4 Invitation to tender

As soon as it has been decided to select a Contractor by competitive tender rather than by negotiation, a shortlist of those considered suitable to be invited to tender should be compiled. Sometimes expressions of interest, interviews and presentations can be considered prior to drawing up the list.

It is critical that all the Contractors on the shortlist have the proven competence, resources, health and safety record and financial stability to carry out the works, so that when tenders are received and assessed, no doubt is in the Employer's mind as to whether a preferred Contractor has the ability and resources to carry out the project. Employers should also review their shortlists periodically so as to exclude firms whose performance has been proven to be unsatisfactory, and also to allow the introduction of new Contractors.

The cost of preparing tenders is a significant element of the overheads of the Employer, the Contractors and in turn their Consultants, Subcontractors and Suppliers, so lists of tenderers should be kept as short as practicable. Dependent on the size and type of project it is recommended that the shortlist should be from four to six names.

11.5 EU Procurement Directives

Whilst not wishing this book to be geography-specific, it is worth some consideration of the detailed European Union (EU) Procurement Directives that apply to all procurement within the public sector, above minimum monetary thresholds which are reviewed on a regular basis, and subject to EU-wide principles of non-discrimination, equal treatment and transparency.

The purpose of the procurement rules is to open up the public procurement market and to ensure the free movement of supplies, services and works within the EU.

The rules are enforced by the Members States' Courts and the European Court of Justice and require that all public procurement must be based on value for money, defined as 'the optimum combination of whole life cost and quality to meet the user's requirement', which should be achieved through competition, unless there are compelling reasons to the contrary.

In addition to the EU Member States, the benefits of the EU public procurement rules apply to a number of other countries outside Europe because of an international agreement negotiated by the World Trade Organization (WTO), entitled the Government Procurement Agreement (GPA).

The EU Procurement Directives set out the legal framework for public procurement when public authorities and utilities seek to acquire supplies, services or works (e.g. civil engineering or building), procedures which must be followed before awarding a contract when its value exceeds set thresholds which are reviewed on a regular basis, unless the contract qualifies for a specific exclusion, for example on grounds of national security.

The Directives require contracting authorities to provide details of their proposed procurement in a prescribed format, which are then published in the *Official Journal of the European Union* (*OJEU*). All companies replying to an *OJEU* advertisement have an equal opportunity to express interest in being considered for tendering. As in all tendering exercises, Employers must ensure that those companies selected to tender receive exactly the same information on which to make their bid.

The notice may be issued electronically and this service may be provided through an intermediary organisation.

Generally, contracts covered by the Regulations must be the subject of a call for competition by publishing a Contract Notice in the *OJEU*.

Choice of procurement procedure

Four award procedures are provided for in the Regulations:

1. *the open procedure*, where all those interested may respond to the advertisement in the *OJEU* by tendering for the contract;
2. *the restricted procedure*, under which a selection is made of those who respond to the advertisement and only they are invited to submit a tender for the contract.
    This allows purchasers to avoid having to deal with an overwhelmingly large number of tenders;
3. *the competitive dialogue procedure*, under which, following an *OJEU* Contract Notice and a selection process, the authority then enters into dialogue with potential bidders to develop one or more suitable solutions for its requirements and on which chosen bidders will be invited to tender; and
4. *the negotiated procedure*, under which a purchaser may select one or more potential bidders with whom to negotiate the terms of the contract. An advertisement in the *OJEU* is usually required but, in certain circumstances, described in the Regulations, no advertisement is required.

The *OJEU* Notice process has associated minimum timeframes for receipt of tenders or requests to participate, dependent on procurement procedure, from the date the Contract Notice is sent.

Stages in the procurement process

The Regulations set out criteria designed to ensure all Suppliers or Contractors established in countries covered by the rules are treated on equal terms, to avoid discrimination on the grounds of origin in a particular Member State.

The criteria cover:

*   *specification stage* – how requirements must be specified;
*   *selection stage* – the rejection or selection of candidates;
*   *award stage* – the award of contract, either on the basis of 'lowest Price' or preferably the 'most economically advantageous tender' to the purchaser.

11.6　Preparing tender documents

The structure of the contract

Title page

The title page is Page 1, which includes the names of the contracting parties and a brief description of the work to be carried out. A list of contents is also included.

There are guidance notes in grey-bordered boxes throughout the Contract Data, but it must be stressed that these notes do not form part of the contract. This fact is also stated on Page 1.

The Contract Data

The Contract Data provides the information required by the conditions of contract specific to a particular contract. Other contracts call this the 'Appendices' or 'Contract Particulars'.

The Contract Data is on Pages 2–9. This is information specific to the contract and, apart from Page 4 (Contractor's Offer and Employer's Acceptance), is required to be completed by the Employer or his representative prior to inviting tenders and before a contract can come into existence. The Employer's Acceptance is completed by the Employer when a Contractor is to be appointed.

The Contractor's Offer is required to be completed by the tendering Contractors and also the pricing column within the Price List. Any information provided by the tendering Contractors may be incorporated into the Works Information, but, as the document states, 'only if the Employer is satisfied that it is required, is part of a complete statement of the Employer's requirements and is consistent with the other parts of the Works Information'.

This is an abbreviated version of the Contract Data in the ECC and includes all terms which occur in italics in the conditions of contract.

The Contract Data includes the following:

- the name and contact details of the Employer;
- the name of the works;
- the location of the site;
- the starting date;
- the completion date;
- the period for reply (weeks);
- the defects date (weeks after Completion);
- the defect correction period (weeks);
- the delay damages (per day);
- the assessment day (date of each month);
- the retention (percentage);
- whether the UK Housing Grants, Construction and Regeneration Act (1996) applies;

- the name of the Adjudicator;
- interest rate on late payment;
- limit of liability;
- insurances (see also Insurance Table – Clause 82.1);
- Adjudicator Nominating Body;
- arbitration details;
- any additional conditions.

The Contractor's Offer is on Page 4.

The information which the Contractor is required to complete is as follows:

- details of the Contractor;
- percentage for overheads and profit added to the Defined Cost for people;
- percentage for overheads and profit added to other Defined Cost;
- offered total of the Prices;
- signature on behalf of the Contractor.

The Employer's Acceptance is also on Page 4.

- The Employer's Acceptance requires a signature by a person authorised to sign on behalf of the Employer.

The Price List is on Page 5 and includes an important guidance note. It is in the form of a traditional bill of quantities but it can be used in a number of different ways (see Introduction).

The Works Information consists of six sections and is on Pages 6 to 8.

These six sections are:

1. detailed description of the works;
2. list of drawings;
3. specifications;
4. constraints on how the Contractor Provides the Works;
5. programme requirements;
6. services and other things provided by the Employer.

The Works Information is normally incorporated by reference to separate documents rather than listing drawing numbers and specification references. If this is the case, the separate documents must be clearly defined. (See below for further details and guidance on Works Information.)

The Site Information is given on Page 9.

As with the Works Information, the Site Information is normally incorporated by reference to separate documents and again those documents must be clearly defined. (See below for further details and guidance on Site Information.)

The Conditions of Contract are in the remaining part of the published document after the Contract Data, on pages numbered CC1 to 12.

The tender documents

The tender documents for an ECSC contract will normally consist of:

- *Invitation to Tender* (including instructions to tenderers on time and place of tender submission).

There is no 'pro forma' version of 'Instructions to Tenderers' within the NEC contracts, but any invitation to tender will give a brief description of the work, who it is for and how tenders may be submitted.

Other inclusions will cover such matters as:

- time, date and place for the delivery of tenders;
- what documents must be included with the tender, including a programme;
- the policy regarding alternative and/or non-compliant bids;
- arrangements for visiting the Site, including contact details;
- rules on non-compliant bids;
- anti-collusion certificate.

Form of Tender

There is no pro forma Form of Tender within the NEC contracts apart from the Contractor's Offer and Employer's Acceptance. This is the tenderer's written offer to execute the work in accordance with the contract documents.

The Form of Tender is normally in the form of a letter with blank spaces for tenderers to insert their name and other particulars, total tender Price and other particulars of their offer. It is essential to have a standard Form of Tender and that all tenderers consistently use the same form, to make assessment and comparison of tenders easier.

11.7 Works Information

Works Information is defined by Clause 11.2(13) as information which either:

- specifies and describes the works or
- states any constraints on how the Contractor Provides the Works

and is either

- in the documents called Works Information or
- in an instruction given in accordance with this contract.

The Works Information describes clear boundaries for the work to be undertaken by the Contractor. It may also outline the Employer's objectives and explain why the work is being undertaken and how it is intended to be used. It says what is to be done (and maybe what is not included) in general terms, but not how to do it or the standards to be achieved. It explains the limits, where the work is to interface with other existing or proposed facilities. It may draw attention to any work or materials to be provided by the Employer or others. It should also emphasise any unusual features of the work or contract which tenderers might otherwise overlook.

This is the document that a tenderer can look to, to gain a broad understanding of the scale and complexity of the job and be able to judge its capacity to undertake it. It is written specifically for each contract. In some respects it is analogous to a shopping list. It should be comprehensive, but it should be made clear that it is not intended to include all the detail which is contained in the drawings, specifications and schedules.

The Works Information describes what the Contractor has to do in terms of scope and standards, and in some cases must not do or include in order to Provide the Works. It may also include the order or sequence in which the works are to be carried out. It also details where the work is to interface with other existing or proposed Contractors or facilities. It will also include any work or facilities or materials to be provided by the Employer or others. It may also include any unusual features of the work which tenderers might otherwise overlook, for example any planning constraints.

The Employer provides the Works Information and refers to it in the Contract Data; if the Contractor provides Works Information, e.g. for his design, this is included in his tender.

The Works Information must be carefully drafted in order to define clearly what is expected of the Contractor in the performance of the contract and therefore included in the quoted tender amount and programme. If the contract does not cover all aspects of the work, either specifically or by implication, that aspect may be deemed to be excluded from the contract.

This is the document from which a tenderer can gain a broad understanding of the scale and complexity of the job and his capacity to undertake it. It is written specifically for each contract.

The Works Information should be:

- clear – unambiguous;
- concise – not overly wordy;
- complete – have nothing missing.

The Works Information is separated into six sections, each including guidance notes in boxes, which are shown in the summary below, and guide the Employer as to what he should insert into the sections when completing the forms prior to inviting tenders.

1. Description of the works

Give a detailed description of what the Contractor is required to do and of any work the Contractor is to design.

A comprehensive, all-embracing description should be considered for the description of the work clause, which should be supplemented by specific detailed requirements.

If reliance is placed solely on a very detailed scope description, an item may be missed from this detailed description and be the subject of later contention.

Where items of Equipment are to be fabricated or manufactured off site by others, it is advisable that the contract sets out the corresponding obligations and liabilities of the respective parties, particularly if these are to form an integral or key part of the completed works.

Contractor's design

Details of any design to be carried out by the Contractor should be given, including the interface between design by the Contractor and design by the Employer.

The default position is that all design will be carried out by the Employer. If any or all of the design is to be carried out by the Contractor, this should be included within the Works Information together with any performance requirements to be met within the Contractor's design. In addition, any warranty requirements and also any future novation requirements should be included within the Works Information.

Any design acceptance procedures should be included, including time scales for submission and acceptance of design.

The Contractor does not start work which the Contractor has designed until the Employer has accepted that the design complies with the Works Information.

Financial accounts

There may be details of financial accounts and records to be kept by the Contractor to prove Defined Cost.

Title

A statement is included of any materials arising from excavation or demolition to which the Contractor will have title (Clause 70.2).

Subcontracting

- Lists of acceptable subcontractors for specific tasks;
- Statement of any work which should not be subcontracted;
- Statement of any work which is required to be subcontracted.

Completion

Completion is defined under Clause 11.2(1), but the Works Information should also define any specific requirements which are required in order for completion to have taken place, for example, and requirement for:

- 'as built' drawings;
- maintenance manuals;
- training documentation;
- test certificates;
- statutory requirements and/or certification.

2. Drawings

List the drawings that apply to this contract.

The Works Information will include drawings which, again, should provide clear details of what the Contractor has to do. Clearly, tenderers must be given sufficient information to enable them to understand what is required and thus submit considered and accurate tenders.

This section could include schematic layouts, plan, elevation and section drawings, detailed working and/or production drawings (if relevant and available).

3. Specifications

List the specifications which apply to this contract.

The Works Information will include the specification, which is a written technical description of the standards and various criteria required for the work, and should complement the drawings. The specification describes the character and quality of materials and workmanship for work to be executed.

It may lay down the order in which various portions of the work are to be executed. As far as possible, it should describe the outcomes required, rather than how to achieve them. It is customary to divide the work into discrete sections or trades (e.g. drainage, concrete, pavements, fencing), with clauses written to cover the materials to be used, the packaging, handling and storage of materials (only if necessary), the method of work to be used (only if necessary), installation criteria, the standards or tests to be satisfied and any specific requirements for completion.

The specification is an integral part of the design. This is often overlooked, with the result that inappropriate or outmoded specifications are selected, or replaced by a few brief notes on the Drawings. The designer should spend an appropriate amount of time specifying the quality of the work, as it is not possible to price, build, test or measure the work correctly unless this is done. This will include:

- Plant and Materials;
- materials and workmanship specifications;
- requirements for delivery and storage;
- future provision of spares and maintenance requirements.

Tests

There should be descriptions of tests to be carried out by the Contractor, the Employer and/or others, including those which must be done before Completion and/or after Completion.

There should be a specification of materials, facilities and samples to be provided by the Contractor and the Employer for tests before or after delivery to the site.

4. Constraints on how the Contractor Provides the Works

State any constraints on the sequence and timing of work and on the methods and conduct of work including the requirement for any work by the Employer.

The Works Information should include a statement of any constraints on how the Contractor Provides the Works, e.g. restrictions on access, sequencing or phasing of works, security issues.

There may also be specific health and safety requirements for the site with which the Contractor must comply, particularly if the site is within existing premises, including house and local safety rules and evacuation procedures.

Any pre-construction information and health and safety plans for the project should be included.

There should be details of any other Contractors and others who will be occupying the site during the contract period and any sharing requirements.

5. Requirements for the programme

State whether a programme is required and, if it is, state what form it is to be in, what information is to be shown on it, when it is to be submitted and when it is to be updated.

State what the use of the works is intended to be at their Completion as defined in Clause 11.2(1).

The Works Information should include any information which the Contractor is required to include in the programme. Also, if the Employer requires the Contractor to produce a certain type of programme or to submit it using a certain brand of software, then this should be clearly detailed within the Works Information.

Any specific submission and acceptance provisions should also be included here.

6. Services and other things provided by the Employer

Describe what the Employer will provide, such as services (including water and electricity) and 'free issue' Plant and Materials and equipment.

The Employer should clearly state what, if anything, he is providing to the Contractor, whether it is 'free issue' or at cost, and if a cost is to be levied, what and how is it to be levied.

11.8 Site Information

Site Information is defined in Clause 11.2(12) as 'information which describes the site and its surroundings and is in the document called "Site Information"'.
The Contract Data states:

> Give information about the site such as the ground conditions and any other information which is likely to affect the Contractor's work such as limitations on access and the position of adjacent structures.

Site Information may include:

(i) ground investigations, borehole and trial pit records and test results. If the Employer has obtained such information, it should not be withheld from the tenderers, though the tenderers should be aware that the Site Information alone cannot be relied upon in terms of a possible compensation event (see Clauses 60.1(9) and 60.2);
(ii) information about existing buildings, structures and plant on or adjacent to the site;
(iii) details of any previously demolished structures and the likelihood of any residual surface and subsurface materials;
(iv) reports obtained by the Employer concerning the physical conditions within the Site or its surroundings. This may include mapping, hydrographical and hydrological information;
(v) all available information on the topography of the site should be made accessible to tenderers, preferably by being shown on the Drawings;
(vi) environmental issues, for example nesting birds and protected species;
(vii) references to publicly available information;
(viii) information from utilities companies and historic records regarding plant, pipes, cables and other services below the surface of the site.

It is vital that care is taken to get the Site Information correct. Tenderers must be given sufficient information to enable them to understand what is required and thus to submit considered and well-priced tenders.

Note that there is no equivalent of ECC Clause 60.3 if there is an ambiguity or inconsistency within the Site Information, whereby the Contractor is assumed to have taken into account the conditions most favourable to doing the work, which would then put the benefit of any doubt in the favour of the Contractor.

In the case of the ECSC each case would be judged on its merits, broadly relying on the rule of *contra proferentem* (*contra* = against, *proferens* = the one bringing forth), in that where a term or part of a contract is ambiguous or inconsistent it is construed strictly against the party which imposes or relies on it.

11.9 Provisional and Prime Cost Sums

Provisional and Prime Cost (PC) Sums are used in other contracts where there are elements of work which are not designed or cannot be sufficiently defined at the time of tender and therefore a sum of money is included by the Employer in the bill of quantities or other pricing document to cover the item. When the item is defined or able to be properly defined, the Contractor is given the information which allows him to price it, the Provisional Sum is omitted and the Price is included in its stead.

The problem with Provisional Sums is that they reduce the competition amongst tenderers as they are not priced at tender stage. Also if they are 'defined' the Contractor is deemed to have allowed time in his programme for them; if they are 'undefined' he has not.

If one considers General Rule 10 of the Standard Method of Measurement of Building Works, Seventh Edition (SMM7), where work cannot be described and given in items in accordance with these rules it shall be given as a Provisional Sum and identified as either 'defined' or 'undefined' work as appropriate.

Rule 10.3 of SMM7 requires that, where the Provisional Sum is for defined work which is not completely designed, the following information shall be provided:

- the nature and construction of the work;
- a statement of how and where the work is fixed to the building and what other is to be fixed thereto;
- a quantity or quantities which indicate the scope and extent of the work;
- any specified limitations and the like identified in Section A35 of SMM7 (limitations imposed by the Employer, e.g. access, disposal of Materials found, working hours).

General Rule 10.4 of SMM7 states that 'where Provisional sums are given for defined work the Contractor will be deemed to have made due allowance in programming, planning and pricing preliminaries'.

General Rule 10.5 states that if the information specified in General Rule 10.3 is not available, the Contract Bills should describe the Provisional Sum as 'undefined'.

General Rule 10.6 states that the Contractor is deemed to have made no such allowance for undefined work.

The NEC3 contracts do not provide for Provisional Sums, the principle being that:

- the Employer decides what he wants at tender stage so it can be designed and accurately described and the tendering Contractors can properly price and programme for them; or
- when the Employer decides what he wants, he can give an instruction to the Contractor which changes the Works Information; this is a compensation event under Clause 60.1(1) and can be priced at the time. In that case, the cost of the work is held in the 'scheme' rather than in the contract.

11.10 Submission of tenders

There is a tendency to give Contractors very short periods of time in which to submit their tenders as the Employer wishes the work to be commenced as soon as possible. However, it is critical that tenderers are given sufficient time in which to consider properly what they have to price, and time and resource-related issues, and to make any enquiries and obtain quotations prior to submitting their tenders.

The time and place for the delivery of tenders need to be clearly stated. It is also critical that all tenderers are treated on an equal basis; therefore it should be made clear to all tenderers that late returns will be invalid and will not be considered.

Late tenders should be returned to the tenderer unopened.

The tender documentation should also make clear the period over which a tender may be required to remain open for acceptance. This would normally be three months, giving the Employer and his advisers time to assess the tenders properly and make the appropriate award.

11.11 Opening tenders

Again, in order to maintain equality between the tenderers, tenders should not be opened until after the date and time for submission to avoid any suspicion of collusion. No tenderer should be allowed to alter his tender after the closing time.

After opening, the tenders and supporting documents should be evaluated.

Each tender is confidential and no details of a tender should be released to any other tenderer.

11.12 Notification of tender results

It is important that all tendering Contractors be informed of the results of tendering as soon as possible after the tender submission date. This applies particularly to those whose tenders are not successful, or which are not being given further consideration, as it frees them to concentrate on opportunities with other Employers.

Generally, Employers' representatives have no authority to accept tenders unilaterally, as ultimately the contract is between the Employer and the Contractor, so it is normal to recommend a choice to the Employer and allow him the final decision and inform the tenderers.

The Employer's representative should submit a report, which lists the tenders received, comments on them and recommends a particular one for acceptance. The scope and detail of this report will vary according to the circumstances. The report should present a clear and logically reasoned case for acceptance of the recommended tender.

The following may be included in the report:

- a tabular statement of all the tenders received, showing the name of each tenderer;
- the tender sum, whether it was a valid submission, and any qualifying conditions;

- reference to any discussions that may have taken place with any tenderers regarding the tender;
- a concise summary of the findings following the examination and assessment of each tender, reasons for considering any tender invalid, any discussion of the programme and methods proposed and reasons for considering any of these to be unsatisfactory;
- comments on any rates or Prices which appear exceptionally high or low, and forecasts of the possible effects on the contract Price in cases where there are large differences;
- a comparison of the tender of the recommended successful bidder with the original cost plan or budget;
- recommendation of the most acceptable tender.

Remember – the cheapest price does not always mean the cheapest cost!

As discussed earlier, by carrying out a detailed pre-tender analysis when tenders are received, the evaluation criteria should, in theory, be fairly straightforward as all the tenders should be capable of being accepted, so it will normally just include any further information gained from the bid itself.

11.13 Awarding the contract

There is no Standard Form of Agreement for completion and signature within the ECSC other than the Contractor's Offer and Employer's Acceptance.

The drafters of the contract have stated that parties enter into contracts in many different ways, such as by executing Forms of Agreement or by exchange of letters. Whilst this is probably true, it would have assisted the parties if there had been a standard Form of Agreement that they could use if they so wish.

The difficulty with not having a Form of Agreement within the contract is that the parties have to write their own. They also have to consider whether the contract is to be executed as a deed, and in that respect they must be very cautious that all the documents that are intended to form the contract are included in the Form of Agreement.

Note

1 M. Latham (1994) *Constructing the Team*. Department of the Environment, London.
2 J. Egan (1998) *Rethinking Construction*. Department of the Environment, London.
3 Construction Industry Institute (CII) Partnering Task Force (1987). Austin, TX: www.construction-institute.org.

Index

Printed in the United States
by Baker & Taylor Publisher Services